Literatures, Cultures,

Ursula K. Heise, St

Literatures, Cultures, and the Environm _____ on new research in the Environmental Humanities, particularly work with a rhetorical or literary dimension. Books in this series seek to explore how ideas of nature and environmental concerns are expressed in different cultural contexts and at different historical moments. They investigate how cultural assumptions and practices as well as social structures and institutions shape conceptions of nature, the natural, species boundaries, uses of plants, animals, and natural resources, the human body in its environmental dimensions, environmental health and illness, and relations between nature and technology. In turn, the series aims to make visible how concepts of nature and forms of environmentalist thought and representation arise from the confluence of a community's ecological and social conditions with its cultural assumptions, perceptions, and institutions. Such assumptions and institutions help to make some environmental crises visible and conceal others, confer social and cultural significance on certain ecological changes and risk scenarios, and shape possible responses to them.

Across a wide range of historical moments and cultural communities, the verbal, visual, and performing arts have helped to give expression to such concerns, but cultural assumptions also underlie legal, medical, religious, technological, and media-based engagements with environmental issues. Books in this series will analyze how literatures and cultures of nature form and dissolve; how cultures map nature, literally and metaphorically; how cultures of nature rooted in particular places develop dimensions beyond that place (e.g., in the virtual realm); and what practical differences such literatures and cultures make for human uses of the environment and for historical reshapings of nature. The core of the series lies in literary and cultural studies, but it also embraces work that reaches out from that core to establish connections to related research in art history, anthropology, communication, history, philosophy, environmental psychology, media studies, and cultural geography.

A great deal of work in the Environmental Humanities to date has focused on the United States and Britain and on the last two centuries. *Literatures, Cultures, and the Environment* seeks to build on new research in these areas, but also and in particular aims to make visible projects that address the relationship between culture and environmentalism from a comparative perspective, or that engage with regions, cultures, or historical moments beyond the modern period in Britain and the United States. The series also includes work that, reaching beyond national and majority cultures, focuses on emergent cultures, subcultures, and minority cultures in their engagements with environmental issues. In some cases, such work was originally written in a language other than English and subsequently translated for publication in

the series, so as to encourage multiple perspectives and intercultural dialogue on environmental issues and their representation.

Ecocriticism and Shakespeare: Reading Ecophobia
by Simon C. Estok

Ecofeminist Approaches to Early Modernity
Edited by Jennifer Munroe and Rebecca Laroche

*Myths of Wilderness in Contemporary Prose Texts:Environmental
Postcolonialism in Australia and Canada*
By Kylie Crane

East Asian Ecocriticisms: A Critical Reader
Edited by Simon C. Estok and Won-Chung Kim

The Environmental Imaginary in Brazilian Poetry and Art
By Malcolm K. McNee

ECOCRITICISM AND SHAKESPEARE

READING ECOPHOBIA

Simon C. Estok

First published in hardcover in 2011 by PALGRAVE MACMILLAN® in the United States—a division of St. Martin's Press LLC, 175 Fifth Avenue, New York, NY 10010.

Where this book is distributed in the UK, Europe and the rest of the world, this is by Palgrave Macmillan, a division of Macmillan Publishers Limited, registered in England, company number 785998, of Houndmills, Basingstoke, Hampshire RG21 6XS.

Palgrave Macmillan is the global academic imprint of the above companies and has companies and representatives throughout the world.

Palgrave® and Macmillan® are registered trademarks in the United States, the United Kingdom, Europe and other countries.

ISBN: 978–1–137–44689–3

The Library of Congress has cataloged the hardcover edition as follows:

Estok, Simon C.
    Ecocriticism and Shakespeare : reading ecophobia / Simon C. Estok.
      p. cm.—(Literatures, cultures, and the environment)
    Includes bibliographical references.
    ISBN 978–0–230–11256–8 (hardback)
      1. Shakespeare, William, 1564–1616—Knowledge—Natural history.
    2. Ecocriticism. 3. Nature in literature. 4. Human ecology in literature.
    I. Title.
PR3039.E88 2011
822.3'3—dc22   2011005270

A catalogue record of the book is available from the British Library.

Design by Newgen Knowledge Works (P) Ltd., Chennai, India.

First PALGRAVE MACMILLAN paperback edition: September 2014

10 9 8 7 6 5 4 3 2 1

for Jonathan, Sophia, and Yeon-hee

CONTENTS

ACKNOWLEDGMENTS

I am indebted to the editorial staff at Palgrave Macmillan (particularly to Joanna Roberts and Brigitte Shull) for their diligence and support, and to the copyediting and production team at Newgen Imaging Systems (particularly to Rohini Krishnan) for their patience and meticulous skill. The many suggestions of an anonymous reader made this a much better book than it was.

This book has dual loyalties—to ecocriticism and to Shakespeare studies. Among the early modern scholars to whom I owe the deepest gratitude are Sharon O'Dair, Lynne Bruckner, Dan Brayton, Karen Raber, Linda Woodbridge, Charles Whitney, and, in particular, the late Lloyd Davis for his courage in accepting what was to become one of the first published pieces of ecocritical Shakespearean scholarship. Among the ecocritics to whom I am for various reasons most indebted are Greta Gaard, Serpil Oppermann, Ursula Heise, Terry Gifford, Patrick Murphy, and Greg Garrard.

I owe a special debt of gratitude to Scott Slovic for giving me a forum to express my theories about ecophobia, to Peter Singer for talking with me about and helping me to work through those theories, to Ruth Ozeki for teaching me about process and praxis, and to Ken Dong for being my friend for so long. I would also like to thank Peter I-Min Huang for graciously letting me use his office (and thus allowing me to put some of the finishing touches on this book) during my month-long visit to Tamkang University in the summer of 2010.

Much of the research for this book was carried out in the Folger Shakespeare Library in Washington, DC. I am deeply grateful to the Reading Room staff at the Folger—Harold Batie, LuEllen DeHaven, Leigh Anne Palmer, Camille Seerattan, and Betsy Walsh—for their immeasurable help, dedication, and professionalism. I would also like to thank here Rebecca Oviedo of the Folger for helping me to get many of the images I needed for this book, and Ellen Charendoff—the archives coordinator for the Stratford Shakespeare Festival (Ontario)—for her relentless persistence in tracking down and delivering to me the John Martin artwork that graces the cover of this book.

I am enormously indebted to the National Research Foundation of Korea for its 2009–2011 three-year grant for the Writing in the Humanities Program (인문저술지원사업—Inmoon Jeosool Jeeweonsa Eop). I wish also to acknowledge the unstinting research support from Sungkyunkwan University, where I have worked very happily since 2007, a position for which I would like to publicly thank Professors Shin Moonsu (Seoul National University), Kim Won-Chung (Sungkyunkwan University), and Kim Jonggab (Konkuk University).

Some of the chapters of this book have been published in various forms elsewhere. An earlier version of chapter 2, "Dramatizing Environmental Fear: *King Lear's* Unpredictable Natural Spaces and Domestic Places," was published in *AUMLA: Journal of the Australasian Universities Language and Literature Association* 103 (May 2005); an earlier form of chapter 3, "*Coriolanus* and Ecocriticism: A Study in Confluent Theorizing," was published in *Shakespeare Review* 44. 3 (Fall 2008); a version of chapter 4, "Pushing the Limits of Ecocriticism: Environment and Social Resistance in 2 *Henry VI* and 2 *Henry IV*," was published in *Shakespeare Review* 40. 3 (Summer 2004); chapter 8, "The Ecocritical Unconscious: Early Modern Sleep as 'Go-Between,'" was first published in *Foreign Language Studies* 30. 5 (Fall 2008); and an earlier version of the "Coda" appeared as the "Afterword: Ecocriticism on the Lip of a Lion" in *Ecocritical Shakespeares*, ed. Lynne Bruckner and Dan Brayton (New York: Ashgate, 2011). Much of the material directly theorizing ecophobia (especially in chapter 1, "Doing Ecocriticism with Shakespeare: An Introduction") first appeared as "Theorizing in a Space of Ambivalent Openness: Ecocriticism and Ecophobia" in *ISLE: Interdisciplinary Studies in Literature and Environment* 16. 2 (2009).

Figure 5.1A, from Sebastian Münster's *Cosmographia* (1554), figure 5.1.B, from Hartmann Schedel's *Liber chronicarum* (1493), figure 5.1.C, from Konrad Lykosthenes's *The doome warning all men to the iudgemente* (1582), figure 5.2, Jan van der Straet's "America," and figure 5.3 (The Cracovia Monster) are used by permission of the Folger Shakespeare Library. The cover illustration is used courtesy of the Stratford Shakespeare Festival Archives.

This work was supported by the Korea Research Foundation Grant funded by the Korean Government (KRF-2009-812-A00133).

This has been a long process, and many people have helped me in many ways, but I owe the deepest gratitude to the people who live with me: my beloved companion, Cho Yeon-hee (who endured several drafts), and my little ones, Jonathan and Sophia (who both learned to walk—sometimes away—while I was writing this book).

1

DOING ECOCRITICISM WITH
SHAKESPEARE: AN INTRODUCTION

When I began this project over a decade ago, I started with questions that seemed much more radical than they seem now. Now that ecocriticism has become established as a credible area of study with increasing insights on Shakespeare,[1] it is less provocative and contentious to suggest that there are radical possibilities that may open up when we apply ecocriticism to Shakespeare. Given what is happening in the field, we can assume now that a Shakespearean ecocriticism is unquestionably useful to contemporary environmental discussions. We can assume that literary theories about representations of environmental issues have a place in serious Shakespearean scholarship; that there is a case (many, perhaps) compelling enough to persuade Shakespeareans of the usefulness of ecocriticism *and* to convince ecocritics that the growth and development of ecocriticism itself stands to gain substantially from readings of Shakespeare; and that applying ecocriticism to Shakespeare is very different from doing thematic nature criticism.

We can also see that the more we talk about representations of nature in Shakespeare, the more clear it becomes that simple green thematicism has become old hat for Shakespeare, that ecocriticism is methodologically committed to confluent theorizing—thus, Breyan Strickler's "Sex and the City: An Ecocritical Perspective on the Place of Gender and Race in *Othello*" locates ecocriticism alongside feminist literary methods and postcolonial insights, displaying a theory that cross-pollinates each area of investigation; the questioning in Sharon O'Dair's "Slow Shakespeare" about requirements of our profession and about how academics overproduce run alongside theoretical matters of historicism and presentism; and Gabriel Egan's ecocritical volleys produce a confluence of discussions about genetics, nuclear fission, and geology, sometimes with quite unexpected results (for instance, that the benzene ring and the Globe Theater seem to have a structural similarity). If, the more we talk about representations of nature, the more clear it becomes that confluent theorizing is necessary, no less has it also become clear that there is a need for permeable borders in ecocriticism.

For better or for worse, central to the ecocritical credo have been commitments to extending the scope and possibilities of ecocriticism, an expansion of the boundaries of ecocriticism to include discussion of "environmental ideas and representations *wherever* they appear" (Kerridge 5). Among the consequences have been accusations of "fuzziness" and subsequent calls for more definitive structure, methodological definition, and viable terminology. Thus, there seems to be a growing need to talk about how contempt for the natural world is a definable and recognizable discourse (what we may call "ecophobia," a term one blogger has perhaps prematurely called a "paradigm"[2]) and how this discourse finds expression in Shakespeare. It is the explicit aim of the pages that follow to provide such discussions.

It is necessary to introduce (and to argue strenuously for) the new term ecophobia because there is simply no appropriate existing term for the concept it seeks to describe. The term opens opportunities to the study of nature in ways similar to the ways terms such as misogyny, racism, homophobia, and anti-Semitism open up studies of the representations of women, race, sexuality, and Jewishness respectively. Without being monolithic and implying that ecophobia defines the primary way that humanity responds to nature (which is certainly not the intent here), the pages that follow identify possibilities for discussions in Shakespeare, discussions that have perhaps been hindered by the lack of an adequate vocabulary. The term ecophobia also takes us to one of the distinguishing birthmarks of ecocriticism: its activist visions.

Part of the radical appeal of ecocriticism in its embryonic stages was its gestures toward activist possibilities, like other "political" theories before it—feminism, queer theory, postcolonial theory, and versions of cultural materialism. Ecocriticism's promise to seek out radical activist possibilities is no doubt what is behind Gabriel Egan's rejection of Jonathan Bate's "claim that ecocriticism should be non- (or in his phrase, pre-) political." Having it so divorced is as absurd, Egan claims, as having nonpolitical Marxist, feminist, postcolonial, and queer criticisms (44). So it is perfectly in line with where ecocriticism has gone for Egan to note that we should "retreat from the blind alley of treating ecocriticism as the study of nature writing" (45) and for him to note further that "just as politicized radical criticism based on gender, race, and sexual orientation takes in the full range of cultural considerations, so Green criticism has an application beyond the obviously green-world plays such as *A Midsummer's Night's Dream*" (175). It seems so obvious, but it needs being said, for nothing comes of nothing.

If ecocriticism is committed to making connections, then it is committed to recognizing that there is a thing called ecophobia and that racism, misogyny, homophobia, and speciesism are thoroughly interwoven with it and with each other and must eventually be looked at together.[3] With Shakespeare, this means looking anew at issues that have already been studied extensively, but with an eye to seeing how they interrelate. It means, for instance, looking at implicit and explicit connections in the texts between human and (nonhuman) animals, the conceptual intertwinings in the drama, and their implications. It means looking at connections and conflicts without being binaristic, a danger that a focus on ecophobia clearly runs (and about which the following pages have much to say), and without being anachronistic. It means reading from a position that acknowledges and theorizes about how our situatedness in the present informs our understandings of the texts, what we choose to take from these texts, and how this situatedness actually enables certain kinds of readings. *Titus Andronicus* is a good case study. Without ecocritical analysis, the play surely will not make meat much of an ethical issue; however, when framed through an ecocritical perspective that shows how the play co-locates the human and the nonhuman on the dinner table, surely something must happen the next time we sit at the table for a meat pie.

This much seems obvious, and, again, sometimes the obvious needs stating. Less obvious, perhaps, is how *2 Henry VI* challenges the ethics of animals as food. At times, so poignant are the representations of the general theme of the ethical intertwining of human and nonhuman animals and so complex are the involvements of the natural physical environment in the drama that it is virtually impossible to avoid a political reading—and it is at these points that the concept of ecophobia becomes vital.

Ecocriticism needs a very broad scope for the term ecophobia.[4] I first proposed the term in 1995[5] rather simplistically "to denote fear and loathing of the environment in much the same way that the term 'homophobia' denotes fear and loathing of gays, lesbians, and bisexuals" ("Reading the 'Other'" 213). David Sobel uses the term to define what he calls "a fear of ecological problems and the natural world. Fear of oil spills, rainforest destruction, whale hunting, acid rain, the ozone hole, and Lyme disease" (5) fall under this category, though Sobel does not go much further than this in defining the term. Clearly, he uses the term differently than I do—for instance, whereas for Sobel, fear of whale hunting *is* (by his definition) ecophobia, I argue that whale hunting is a *result* of ecophobia, of a generalized fear or contempt for the natural world and its inhabitants.

Clinical psychology uses the same term to designate an irrational fear of home; in ecocriticism, the term is independent of and in no way derived from the manner in which it is used in psychology and psychiatry.

In 1999, Robin van Tine proposed a similar term—"gaeaphobia"— (independently, it seems, since there are no references to his source for the term), which he defines as "a form of insanity characterized by extreme destructive behavior towards the natural environment and a pathological denial of the effects of that destructive behavior" ("Gaeaphobia" web). Potentially useful though it is for its identification (sometimes quite mechanical) of attitudes toward the natural environment in terms of pathologies laid out in the *Diagnostic and Statistical Manual of Mental Disorders (DSM IV)*, van Tine's article has not been referenced in any scholarship anywhere that I can find. While this is a bit distressing, van Tine's scholarship is important nevertheless because it shows that the kind of theoretical articulation I am seeking in defining ecophobia has been recognized as being necessary in the field of ecopsychology. My approach, while it does not reject ecopsychological analyses of the pathologies behind contempt for the natural environment, is more interested in the confluent approach that examines philosophical underpinnings.

Broadly speaking, we may define ecophobia as an irrational and groundless fear or hatred of the natural world, as present and subtle in our daily lives and literature as homophobia and racism and sexism.[6] It plays out in many spheres; it sustains the personal hygiene and cosmetics industries (which cite nature's "flaws" and "blemishes" as objects of their work); it supports city sanitation boards that issue fines seeking to keep out "pests" and "vermin" associated in municipal mentalities with long grass; it keeps beauticians and barbers in business; it is behind landscaped gardens and trimmed poodles in women's handbags on the Seoul subway system; it is about power and control; it is what makes looting and plundering of animal and nonanimal resources possible. Self-starvation and self-mutilation imply ecophobia no less than lynching implies racism. If ecocriticism is committed to making connections, then it is committed to recognizing that control of the natural environment, understood as a god-given right in Western culture, implies ecophobia, just as the use of African slaves implies racism, as rape implies misogyny, as "fag-bashing" implies homophobia, and as animal exploitation implies speciesism.[7]

Theorizing ecophobia[8] does not mean offering a new perspective, one that ecocritics have somehow missed; of course, ecocritics have long theorized on matters of anthropocentrism. Theorizing ecophobia does not dismiss but rather builds on that history, offering a vocabulary that is new, a vocabulary for conceptualizing something we do (and have been doing for a long time) and for which we haven't had appropriate descriptive or theoretical words. This is a book as much about theorizing ecophobia as it is about reading Shakespeare through ecocritical lenses.

Theorizing ecophobia means recognizing the importance of control. One of the constitutional moments in Western history has control as its key issue: the biblical imperative about human relations with nature gives Man (a man, actually: Adam) divine authority to control everything that lives. Ironically, the more control we seem to have over the natural environment, the less we actually have. As Neil Levy so aptly puts it, "We are not in control of the non-human world, because we are unable to predict with any accuracy the effects of our actions upon it" (210). Increasingly, the effects of our actions are more intense and less predictable, producing in turn, though, a very predictable storm of ecophobic rhetoric. It is the unpredictability that seems key in so much of what is so integral to "nature" in Shakespeare, whether it is the imagined unpredictability of sexual minorities (as I will argue *Coriolanus* to be), of the witches, of Moors, of women—in each case, matters of power are contingent on assumptions of predictability, and in each case, "nature" is fused with all of the fear and loathing that results when imagined unpredictability prevails in the drama. Discussions of ecophobia take us into the heart of these matters and give added insight to discussions that have already been made but that have ignored the natural environment.

Imagining badness in nature and marketing that imagination—in short, writing ecophobia—is such a multifaceted affair that it is difficult to know where to begin. To an audience such as the Elizabethans, who were very familiar with grain shortages, bad harvests, cold weather, and profound storms, we may easily see how someone such as Shakespeare writes ecophobia in a play such as *King Lear*. Although the play does more than simply engender and reflect fear, as I will show, it is vivid in its foregrounding of environmental unpredictability, its dramatization of a king powerless before nature, of a king who is victimized by the weather, unhoused, and alienated. To a global audience glued before flatscreens of CNN, an audience very familiar with polar icesheets breaking off, global warming, and devastating hurricanes, we may easily see how our media daily writes nature as a hostile opponent who is responding angrily to our incursions and actions, an opponent to be feared and, with any luck, controlled.

Human history is often a history of controlling the natural environ-ment, of taking rocks and making them tools or weapons to modify or to kill parts of the natural environment, of building shelters to pro-tect us from weather and predators, of maintaining personal hygiene to protect ourselves from diseases and parasites that can kill us, and of first imagining agency and intent in nature and then quashing that imagined agency and intent.

Nature often then becomes the hateful object in need of our con-trol, the loathed and feared thing that can only result in tragedy if left in control (as in *King Lear* or Hurricane Katrina). Control of nature means arguing on the biblical precedent, as Francis Bacon does, that nature exists for and because of mankind, that "Man, if we look to final causes, may be regarded as the centre of the world; insomuch that if man were taken away from the world, the rest would seem all astray, without aim or purpose" (Bacon 270). As Keith Thomas has noted, "[F]or Bacon, the purpose of science was to restore to man that dominion over the creation which he had partly lost at the fall" (27). This is a far cry from the more recent positions that evalu-ate our relevance—Christopher Manes, for instance, arguing that "If fungus, one of the 'lowliest' of forms on a humanistic scale of values, were to go extinct tomorrow, the effect on the rest of the biosphere would be catastrophic; in contrast, if *Homo sapiens* disappeared, the event would go virtually unnoticed by the vast majority of Earth's life forms" (24).[9] If such "ecological humility," as Manes terms it (17), is one of the hallmarks of ecocriticism, though, ecophobia is one of the hallmarks of human "progress."

Theorizing ecophobia requires at least some discussion of its history,[10] sketchy though it must be here, my purpose—given space limitations—being less to provide a detailed historical analysis than the briefest of outlines. While ecophobia has one of its most famous articulations in the Old Testament, it does not begin there. It proba-bly has roots that reach back to the evolution of the opposable thumb, which enabled hominids to make tools and to conscript "wheat, bar-ley, peas, lentils, donkeys, sheep, pigs, and goats about 9,000 years ago" (Crosby 21). By the early modern period, obviously, there had been huge changes in humanity's relationship with the natural world, and, without a doubt, the crossing of the seas in the fifteenth century and the subsequent empire-building that developed produced the most dramatic of those historical changes up to that point. Imperialism indirectly offered the first big push to control of the natural environment since the Neolithic Revolution. The world was becoming smaller, mappable, predictable, and less diversified. With the colonists came disease, extinctions, homogenization, and profound changes in humanity's control of the world. The romanticization of nature as a space of simplicity, innocence, and peace that Raymond Williams notes as characteristic of "the country" no more slowed the progress of ecophobia than did the notion of "the Noble Savage" slow the genocide of colonized peoples in the New World.

Not far behind the crossing of the seas and the colonialism that developed forthwith was, of course, the Industrial Revolution. Here, the control of nature was consolidated. Among the many paradigmatic shifts and lurches occasioned by the Industrial Revolution was the redefinition of nature from participative subject and organism in an organic community to a pure object, a machine that ideally could be intimately and infinitely controlled and forced to spit out products in the service of an increasingly utilitarian capitalist economy.

Though we can always find diggers and levelers and pockets of resistance that challenge the ecophobic hegemony of the West, and though there is undeniably a biophilic impulse somewhere inside humanity, history has not been kind to green thinkers and revisionists. Even now at what seems a new height of interest in environmental issues, we continue to hear a pathological inability to see connections in the language of ecophobia, the labeling—for instance—of the natural world as an angered Mother Nature; we continue to see the versatility of ecophobic stances that posit nature as the scapegoat for social problems; and we continue to see people who fashion themselves as part of the solution actively resisting the kinds of theory that might indeed help lead to answers that have been so persistently out of our reach.[11] Embedded in a history that has regularly been ecophobic, Shakespeare requires an ecocritical lens that clarifies how his representations of nature often, though clearly not always, function.

Doing ecocriticism with Shakespeare means doing it with an eye to what was going on at the time, and, as Sharon O'Dair has argued convincingly that "ecocriticism of Shakespeare is presentist" ("Is It Shakespearean Ecocriticism If It Isn't Presentist?" 85), ecocritical Shakespeares must come "from," to quote Kathleen McLuskie, "the perceived correlation (or its lack) between the plays and contemporary pre-occupations" (239). Of course, we need not necessarily look for "Shakespeare dramatizing problems that were embryonic in early modern England," as Kiernan Ryan puts it (171). Presentism is not necessarily about detailing current environmental issues in a linear progression from their past to current manifestations. Ryan's remarkably clear discussion of presentism notes that "[t]he greatest strength of presentism is its recognition that the present is the place from which critics must start any encounter with Shakespeare's works" (173). This seems straightforward enough. Indeed, presentism is a fairly simple concept, one that many of us have long practiced in our work in feminist research, for instance. It is perhaps surprising, therefore, to find entire monographs being written on the topic; nevertheless, because presentism is obviously very important to Shakespearean ecocriticism, it is entirely appropriate to have presentist concerns frame and run through this book, sounding, at the same time, a few of presentism's salient features.

Hugh Grady, certainly one of the leading voices articulating presentist concerns, has noted that "whether the past is constructed as Other or as our own, it is always defined by its relation to ourselves and our self-understanding." Grady goes on to assert that "there can be no historicism without a latent presentism, [and] that our attitude towards the past governs our approach to the artifacts of that past, and that attitude is a function of our present specific social and historical situation" ("*Hamlet* and the Present" 143). If we assume (as I do) that the bottom line with activist criticism of any sort is that it has got to cause changes not only in how we think but also in how we interact with the material world, then presentism becomes central to the ecocritical Shakespearean project. Doing ecocriticism (implicitly activist) with Shakespeare will invariably mean recognizing the importance of discursive pasts for material presents. It will mean recognizing the importance of how present concerns shape our inquiries into the past, how, to borrow again from Grady, "all our knowledge of the past, including that of Shakespeare's historical context, is shaped by the ideologies and discourses of our cultural present" ("Why Presentism Now" web). And it will mean doing this carefully, with Ryan's warning about the perils of presentism in mind: "Omnivorous assimilation—the complete colonization of then by now—is the peril," Ryan warns, "that presentism must avoid at all costs" (173).

Doing ecocritical Shakespeare is a difficult business, one very different in many ways from doing ecocriticism with someone such as Thoreau. With Shakespeare, doing ecocriticism is something of a balancing act between valid Shakespearean scholarship on the one hand and real ecological advocacy on the other. The fact that many

twentieth- and twenty-first-century "nature writers" are often explicitly political and direct in their comments about nature is perhaps
what has led ecocriticism away from self-theorizing (which has not
been needed) and away from articulating any kind of uniquely ecocritical methodology (which has also not been needed); moreover,
the temporal proximity of the analytical subject of contemporary ecocriticism (much written in the twentieth- and twenty-first centuries)
guarantees that the kinds of questions between history and present
are far less relevant than is the case with ecocritical Shakespeares.

This work clearly requires tools that are new. Ecocriticism, a family of new tools, has different branches (much as screwdrivers have
different varieties with very clearly different purposes and abilities).
Theorizing ecophobia, one of the branches of ecocriticism, helps us to
articulate a methodology, and I will demonstrate below just how useful and necessary such theorizing is. It is not intended as a monolithic
theory that has all of the answers. It is not the universal screwdriver.

Doing ecocriticism with Shakespeare, despite the volume of material currently appearing, still faces skepticism. In a seminar in the
summer of 2005, Greg Garrard noted as "limitations" in his "Green
Shakespeares" talk the fact that "most of [Shakespeare's] plays do not
deal with the natural world or animals in any significant way, and the
historical context in which he wrote was neither afflicted by major
environmental problems, nor plagued by doubts about the role of
humanity on Earth." This, however, seems inaccurate, since there
were indeed many environmental problems at the time; pollution,
perhaps, being the biggest issue.

William Harrison's unease, for example, about air pollution and
how it might affect people's health was a growing concern of many
people in post-feudal England. The pollution problem became so serious that by the Restoration, treatises on the subject were beginning
to reach the Crown.[12] The air was especially bad in the cities. London
air was filthy, largely the result of coal burning and deforestation.[13]
Long before the Restoration, though—in fact, as early as 1543, with
Henry VIII's *Act for the Preservation of Woods*—there were concerns
about the state of England's forests.[14] And we know that sixteenth
and seventeenth-century England required timber for a variety of
industrial and domestic purposes: firewood for the construction of
houses and other buildings, furniture, household utensils, carts, wag-
ons, posts, rails for fencing, hurdles, troughs, dairy utensils, tool han-
dles, glass-making (until 1650, after which time coal was used), fuel
to heat dyeing vats for the garment industry and brine for the production of salt, iron-smelting, and ship-building, as N. D. G. James

has noted in *A History of English Forestry*. At the end of the sixteenth century, John Manwood was writing that "the greatest part of [forests] are spoiled and decayed" (*2), and by the seventeenth century, the concerns had become quite serious. Whole texts were appearing at this time concerned entirely with the problem. Arthur Standish, writing in 1611, complained that "few or none at all doth plant or preserve" ($C_1{}^R$) and offered remedies for the increasing problem of supply. In the following year, Rooke Churche argued for the need "to propagate and advance this most needfull thrift of planting young trees, and sowing of their seeds (where scarce of wood is) for increase of timber and fuell" ($A_4{}^R$). The main issue was supply. James contends that there were two reasons for the problem of supply: first, there were logistical problems with transportation of the materials; second, there had been a "lack of adequate steps in the past to ensure a supply for the future" (124). According to James, "much of the concern for possible timber shortages stemmed from the fear that supplies for the Navy might be jeopardized" (122), which raises questions about the relationship between conservation and imperialism,[15] which in turn prompts us to think about the ethics behind the early modern preservationist movements.[16] There were laws, but, as James points out, they were difficult to enforce, and many exceptions were made—both of which, in effect, left the laws toothless.

Though there were, of course, major environmental problems before, during, and after the time of Shakespeare, the final question of Garrard's 2005 paper was and remains provocative: what are the possibilities for a "green Shakespeare"?

The ground really begins to open up with Bruce Boehrer's fascinating *Shakespeare Among the Animals*, the conclusion of which gestures promisingly toward ecocritical Shakespeares, predicting that "the ecocritical project will inevitably, and rightly, inform critical responses to [his] book" (181). Gabriel Egan's *Green Shakespeare* (2006) seeks to be more consciously "ecocritical" from the start, to make explicit links with ecocriticism, and to be "political" (44); however, there are enormous gaps that Egan's book leaves open. Indeed, the book virtually ignores ecocriticism. It barely cites ecocritics, reading as though it were trailblazing in totally uncharted territory; but "Shakespeare and ecocriticism" is *not* uncharted territory. Several chapters had appeared in books, several articles in journals, and an entire "special cluster" in *ISLE* had appeared, none of which made it into the book; all were available before Hurricane Katrina, which *did* make it.[17] Terry Gifford's review of the book is a witty insight on the general consensus among ecocritics about the further possibilities for "green Shakespeares": "many critics might have thought," Gifford suggests, "they would like to write a book of this title some day. Well, there is still time, if not a title" (Gifford 272).

Ecocriticism fills gaps, but Egan's book seems rather to open gaps. To the extent, for instance, that it deals with animals from an activist position, the book is concerned with the ethics of animal rights and animal liberation rather than with connections between animals and environmental ethics, arguing that in Shakespeare "human society is not so different from animal society" (102), "that we have much in common with animals" (107), and that "the more we discover about animals, the harder it is to maintain the distinctions between them and us that have become so firmly entrenched since Shakespeare's time" (174). True though some of these anthropocentric statements may be, they are not ecocriticism. Ecocritical activism must go further than simply recognizing continuities between human and nonhuman animals, and it must go further than "animal rights" or "animal liberation" (though these are clearly related issues and are not opposed to ecocriticism).

Although Robert Watson's *Back to Nature: The Green and the Real in the Late Renaissance* (also published in 2006) is a much more sophisticated and scholarly book in many ways, it falls into the same trap as Egan's book of choosing one of its two topics (late Renaissance) over the other (ecological advocacy, which Watson in many ways equates with ecocriticism). Watson's book opens with the promise that it will bring "ecological advocacy into the realm of Renaissance literature" (3), yet it seems in many ways hostile to such a project. It is more than the hostile questions the book asks of ecocriticism and environmental advocacy that makes this so, yet such hostile questions also need attention and are worth quoting at length. Watson asks,

> Is ecocriticism—like New Historicism, some might argue—mostly an effort of liberal academics to assuage their student-day consciences (and their current radical students) about their retreat into aesthetics and detached professionalism, by forcing literary criticism into a sterile hybrid with social activism? . . . is ecocriticism the latest resort of identity politics in the academy, a way for those excluded by the usual categories to claim victim status, either by identifying with an oppressed biosphere . . . or else by imagining their suffering and extinction in an anticipated ecological catastrophe? (4)

Watson's questions raise issues that have dogged recent theoretical approaches, especially those with apparent activist inspiration, as if these issues are the "natural" legacy of ecocriticism's turn to the historical. But his application of these same arguments exposes his underlying assumption that environmentalist movements represent merely a "search for a politically safe and aesthetically attractive version of late 1960s radicalism" (5). Watson's fear that environmental concerns are a "search" for window-display activism is, given the real and violent depredations that ecocritical readings seek to expose, misplaced, and that aspect of his work, combined with the lack of reference to core and basic ecocritical texts, not to mention ecocritical Shakespeares, can give activist ecocritics the sense that Watson is perhaps more the early modernist than the ecocritic. Erudite in the late Renaissance, Watson's book, like Egan's, fails at fully comprehending either the history or the goals of *ecocritical* theory, however excellent his understanding of other theoretical fields may be.

Watson refers (without any apparent intended irony or critique) to "modern nature-lovers" (32) echoing the belittling and dismissive term "animal-lovers" used by detractors of animal rights. Watson's use of the term "nature-lovers" is consistent with the antiecocritical tone the author seems to establish from the beginning of the book. Using the term "nature-lovers" is inappropriate in a book that claims to do ecocriticism: ecocriticism is no more about schmaltzy appeals for the cuteness of animals or the loveliness of nature than animal rights is about sentimentalism (or inordinate love) for animals.[18]

Yet, while both Egan and Watson fail either to follow or to articulate an ecocritical methodology or to advance ecocriticism in theoretical terms, the gaps that they each in their very different ways reveal provide important starting points for this book. The pages that follow will offer nuanced and developed close-readings of Shakespearean drama using a methodology that is both encompassing and focused. The approach encompasses feminism, queer theory, critical racial theory, food studies, cultural anthropology, ecopsychology, poststructuralism, and deconstruction and demonstrates just how productive a theoretically informed historicist ecocriticism can be. The pages that follow below seek to fill in some of the gaps created by recent attempts to do ecocritical readings of Shakespeare, to follow the paradigm of "ecophobia" as it is expressed in Shakespearean drama, and to define clearly the need for this paradigm within and beyond Shakespearean scholarship.

Chapter 2 begins the discussion and builds on much of the very useful work done on nature in *King Lear*. Deviating radically, though, from John Danby's assertion that the play presents nature as either good or bad (but definitely one or the other), chapter 2 argues that neither the play nor the audience to which it was performed see

DOING ECOCRITICISM WITH SHAKESPEARE   13

nature as either one or the other. There is abundant evidence that in Shakespeare's England, in fact, "nature" was an extremely complex site of signification, not reducible either to Danby's good/bad dichotomy or to Robert Watson's "crisis of origins" argument. *Lear* argues not that nature is good or bad or edenic but that it is unpredictable. The fear of nature's perceived or imagined unpredictability (a fear which is constitutive of ecophobia) within the play dramatizes an imagined natural world whose patterning of power relationships, identity, and the notion of home present that natural world as an antagonist. Ecophobia is pervasive in the play and is easy enough to catalog, and the conservative warnings the play offers about what tragedies happen when nature goes unbounded are equally clear. What is less clear is why weather and environment must play so central a role in this tragedy. Theories of analogical thinking do not really answer so much as restate this question. Among the other things chapter 2 does is to look at the actual weather of Shakespeare's England and its relationship with things such as economic growth, with early modern notions about witchcraft, and with the obviously changing views toward nature (from organicism to mechanism).[19] Data are increasing on the colder temperatures, crop failures, and poor fishing that were so very much a part of the early modern daily realities. Chapter 2 thus begins a case that the whole book pursues for a vocabulary from within ecocritical theory for the environmental ethics and attitudes of a play such as *King Lear* and moves beyond the thematic and symbolic readings that have until very recently been most characteristic of the work on Shakespeare's representations of nature.

Chapter 3 posits novel investigations into *Coriolanus*, less through thick descriptions of the weather (which, data shows abundantly, perhaps accounts for the famine that begins the play); rather this chapter looks at relationships between ecophobia, displacement, and voice in the play. While Janet Adelman is accurate in arguing that *Macbeth* and *Coriolanus* effectively reach different conclusions on the relationship between the powerful mother and autonomous masculinity in each play, deep complications of this argument emerge from ecocritical perspectives. *Coriolanus*—embroiled in debates about voice, sexuality, and place—posits a crisis of identity as a crisis of environmental embeddedness and thus demands a kind of confluent theorizing to which ecocriticism is particularly suited. Theorizing ecophobia—because it is inherently predisposed to confluent theorizing and thus to connecting with other "political" theories—may very well in fact lead to the kinds of methodological and structural definition some ecocritics seek. This is partly because the methodology of the ecocritical project begins with analyses of texts—literary and nonliterary—not only through nuanced discussions of the cultural and intellectual history surrounding a given text but also from an extension of such discussions to the environmental history and environmental circumstances surrounding that text.

For a text such as *Coriolanus*, which enunciates radical comments about sexual mobility and about spaces of nature, ecocriticism offers an organizational theoretical framework that makes sense of matters often discussed thematically as unrelated issues; these issues need to be discussed in confluence. So doing allows us to see that Coriolanus, on whose body is mapped an ethically inconsiderable environmental schema, not only becomes subject to the same handling as nature but also becomes constitutionally indistinguishable from it. Coriolanus, in seeking to separate himself from his society, becomes indistinguishable from the natural world but like a weed or a disease that must be cut away, in a space that is no space, the space of same-sex love, a loathed and feared no-man's land, as it were, somewhere between heterosexual marriage and same-sex friendship, between Rome and Corioles, a space that, in this play, cannot be inhabited or voiced. Such work is necessary because through it we can discuss the early modern triumph of individualism in terms of the linked dynamics of what were to become homophobia and ecophobia, dynamics which were becoming increasingly intense during the period.

Committed to an activist stance, ecocriticism works well with texts that explicitly deal with relationships among the representations of social unrest, disease, and environment. Using *2 Henry VI* and *2 Henry IV* as the primary texts, chapter 4 looks at relationships among social resistance, environmental ethics, and matters of disease. One of the issues that arises is vegetarianism. This chapter discusses both the stage marginalization of vegetarianism and its "real life" early modern proponents, showing along the way at least some contemporary implications to the debates that were so clearly raging at the time. Another issue that this chapter discusses is the matter of illness as it relates to the environment in Shakespeare's England. From an ecocritical perspective, precious little has been done on this topic. Chapter 4 thus shows how a variety of class matters are structured and referenced by thoughts about nature; and how chaos and nature, disease and social unrest, environmental ethics and community are importantly connected. It is this *kind* of connecting for which ecocriticism is so well suited.

Monstrosity is the topic of chapter 5. It is time for ecocriticism to talk about monsters.[20] Chapter 5 begins on the assumptions that ecophobia is central to the early modern imagining of hostile geographies; that such geographies house truculent, disenfranchised, and monstrous figures; and that such figures are on (and are often synecdochal of) the outside borders of decency and order. This chapter discusses postcolonial ecocriticism, arguing that the semiotics of cannibalism is one of the vitally overlapping areas between postcolonial theory and ecocriticism. Cannibalism is a race and environment issue. As with so many other of the topics in this book, cannibalism—a matter very present in both the theater and the popular imagination of early modern England—is almost never a topic of ecocritical consideration. One of the things this chapter does is to summarize the important work done in postcolonial studies on the cannibal question as it relates to and is important for ecocriticism, and how xenophobia and racism reciprocate the enabling and encouragement ecophobia proffers. This chapter also, however, argues that "metaphor"—the conceptual links it forces, its organizational function, and its inescapability—plays a central role in the scripting of new world dreams and old world nightmares; if the cannibal is an excitingly exotic new world figure, it is no less a horrifying old world locus of incest and inbreeding. The metaphor of cannibalism in *Pericles* is worthy of ecocritical attention because it dislocates the human into the category of the edible natural world, and, like *Lear*'s "monsters of the deep" that prey upon themselves, *Pericles* presents a natural world that is contemptible and horrible. The metaphors of cannibalism and the frequent association of incest with nature gone wrong in a sense displace responsibility from the human to the wayward natural, exonerating the man from child abuse. This chapter picks carefully through the implications of the cannibal metaphor, examines the metaphors early modern writings use about incest (and there are many such writings), suggests ways that these things contribute strongly to ecophobic fashionings of the natural world, and argues (among other things) that because the semiotics of cannibalism has changed very little over the past four hundred years (and I provide substantial evidence), there is continuing practical relevance for the kind of theoretical comments this chapter makes. This chapter assumes that the "constitutive outside" (Butler 8) of humanity that Judith Butler speaks of rests on and reinforces a foundation of fear and loathing toward "nature" (floral, faunal, geographical, and meteorological) that is profound, boundary work well-addressed through theorizing ecophobia.

Chapter 6 looks at how topics such as disgust, pollution, and gender form a kind of nexus in Shakespeare and both require and benefit from ecocritical readings—readings which show again important links among ecophobia, misogyny, racism, and the persecution of various kinds of social minorities. Because so much of this linking is metaphorical, this chapter goes into an extended discussion about theories of metaphor and about exactly *what* metaphor can carry across. The chapter finishes with a case study of *The Winter's Tale* and applies the theory to the text, suggesting in the process that cross-breeding is definitional to the ideas about pollution and rot on which the text depends; that the spatial dimensions of the play are inseparable from the play's environmental ethics; and that this play expresses profound ecophobic anxieties about the control of nature (a control that the play figures as inseparable from men's control of women).

Chapter 7 argues that with the growing map of uncharted and unknown places in the early modern period, the opening of vast new worlds of resources—natural and human—that had to be controlled before they could be hocked as commodities, ecophobia plays a central role in the environmentally hostile imagination that sought control and domination. Ecocriticism gives access to this process, to understanding *how* ecophobia works, *how* it helps to write geographical and social difference, and to seeing what sorts of relationships we might expect between historical contexts and the staging of ecophobia as well as between the connections ecocriticism and postcolonial theory are currently developing with each other. Discourses of madness form a foundational base on which much Shakespearean exoticism and otherness is grounded. The commodification of this exoticism in characters as varied as Caliban, Shylock, Portia, and Antonio reveal an interdependence of oppressions, each contingent in their varying ways, on ecophobic ethics.

Finally, chapter 8 approaches a topic completely new, it seems, to ecocriticism: representations of sleep and the implications of these in the early modern mind. Again, using Shakespearean drama as the primary grounding of these discussions, this chapter draws on a rich canon of early modern writings about sleep and shows that sleep represents an interstitial space between "the human" and "Nature" in the early modern imagination and that the contempt for both sleep and night are inseparable from a generalized contempt for the natural world.

The coda argues that ecocriticism cannot remain the free-for-all that it has often been, that it must be committed to activism in ways that other "political" theories have been and in ways that thematic criticism has not, and that it must connect with Shakespeare in ways satisfying to both Shakespeareans and ecocritics alike. Certainly, one of the "activist" goals of ecocriticism (like feminism before it) is in articulating what is going on in various power relationships, in having the terms to describe the dynamics, and in making connections with other disempowering relationships. In Shakespeare, this means much more than counting clusters of images and themes.

Doing ecocriticism with Shakespeare is hard work. Ecophobia provides a much-needed paradigm with which to discuss and theorize relationships, and it is a paradigm eminently applicable to Shakespeare (but no less to Thoreau). Doing ecocriticism with Shakespeare extends huge amounts of foundational work that has already been done into exciting new areas, helping us to understand where we have come from, where we are, and where we might be going. Refreshingly, it forces us, in some ways, back to the radical possibilities with which the embryonic ecocriticism all began and gives us new insights and perspectives on a dramatist who indeed had a lot to say about the natural world.

2

DRAMATIZING ENVIRONMENTAL FEAR: *KING LEAR*'S UNPREDICTABLE NATURAL SPACES AND DOMESTIC PLACES

King Lear is vivid in its foregrounding of environmental unpredictability and in its dramatization of a fear of nature. The play markets this dramatic ecophobia to an audience very familiar with grain shortages, bad harvests, cold weather, and profound storms. It was a time of unprecedented exploration, perhaps in part owing to the poor harvests and lack of local fish,[1] and the world was getting smaller. The control of that world and of nature was getting much more desirable and attainable, yet here is Lear powerless within his own kingdom, victimized by the weather, unhoused, and alienated. Ecocriticism offers to give a vocabulary to the environmental ethics and attitudes of *King Lear* and to move beyond the thematicism and symbolic readings that have characterized so much of the critical work on Shakespeare's representations of nature. Ecocriticism helps both to make sense of the startling fear of environmental unpredictability the play presents and to contextualize this ecophobia.

By far the most extensive study of nature in *King Lear* remains John Danby's *Shakespeare's Doctrine of Nature*, a remarkable fact, given the central role of nature in this play. Danby proposes that the play offers a binary vision of nature with a third position in the middle. He locates Lear, Gloucester, Albany, and Kent on one side and characterizes this as the "orthodox" view in which nature is orderly, benign (but punitive), and connected with custom, reason, and religion. On the other side are Edmund, Cornwall, Goneril, and Regan, who are associated with a nature that is at best indifferent to social order and customs and at worst amoral and rapacious. In the middle is Cordelia, whom Danby sees as "standing for Nature herself" (20). For Danby, "Cordelia expresses the utopian intention of Shakespeare's art" (126). Such an approach is useful for understanding authorial intentionality in the play and for offering ways of associating character types with imagined kinds of nature; however, it also argues that characters (such as Edmund) can be outside of nature.

Danby's work, of course, predates ecocriticism and neither seeks nor offers the kinds of ecocentric readings that scholars are increasingly attempting. Ecocriticism, very far removed from the kind of theories Danby offers, spurns both the binaristic thinking that undergirds so much of Danby's argument as well as the idea that anyone is ever outside of nature. Indeed, it is the very inescapability of nature, of characters being subject to its unpredictable and inexplicable power, that produces so very much of the suffering in *King Lear*.

Weather was inexplicable, and Shakespeare's contemporaries were desperate for answers, for scapegoats, and for blood. Women accused of witchcraft and of influencing the weather paid with their lives (see Behringer 335–51; Fagan 91; and Oster 215–28). It was, of course, well within the realm of acceptable thought that mere mortals could, in fact, control the weather. Lear, however, is a king singularly without control, is no Prospero, and has, as he well knows, no control over the environment: "you owe me no subscription," he says, continuing, "Here I stand your slave" (3.2.18, 19).[2] In his masochistic ranting to the storming skies, he commands that the elements do what they are doing anyway.

If ecocriticism is a new source of insight here, then how can we calibrate relationships among the play's representations of very present hostile natural environments, of disaccommodations into those environments, and of the omnipresent nothingness that initiates the tragic action, runs through the play, and dominates so much of the *Lear* critical commentary? Indeed, what is going on with the play's dual obsession with weather and space? On a very basic level in the classroom, these are among the first issues that come up from students.

One of the things an ecocritical reading brings out is that the question of power in *King Lear* has very broad social and environmental implications and that the dependence of identity on environmental control is very strongly influenced by weather in this play.

The Little Ice Age[3] had done a lot to dislocate humanity from its imagined role of authority and control. "Storm activity had increased by 85 percent in the second half of the sixteenth century," according to Brian Fagan (91). Describing the now well-documented Little

Ice Age, he argues that "throughout Europe, the years from 1560 to 1600 were cooler and stormier, with later wine harvests and considerably stronger winds than those of the twentieth century" (90); "the weather had become decidedly more unpredictable, with sudden shifts and lower temperatures that culminated in the cold decades of the late sixteenth century" (xvi); and "as climate conditions deteriorated, a lethal mix of misfortunes descended on a growing European population" (91). Perhaps it is not so far wrong to agree with Stephen Greenblatt that for this play's central concerns, "Shakespeare simply looked around him at the everyday world" (*Will* 357). While Greenblatt does not actually make mention of the materially present weather ruffling and cooling Shakespeare's diminishing hair and years, the intense foregrounding and characterization of weather in *Lear* certainly bear material implications pertinent to the period. As Robert Markley has so eloquently put it,

> Lear is not wandering through a metaphoric storm that marks his poetic madness and signals the disruption of the natural order but an all-too recognizable figure who registers the complex connections between climactic instability and its potential consequences: the loss of agricultural harvests and the fracturing of ideologies of national unity, patriarchal authority, and socioeconomic stability. (137)

Although John Danby is no doubt correct in singling out Edmund as The New Man, the individualist arrogantly seeking profit and self-advancement in a quickly evolving capitalist economy, Lear is no less a single figure fighting alone against what becomes for him a very materially difficult world. Significantly, much of that world that he is fighting against is nature, and, of course, the other thing is that, unlike Edmund, Lear loses—from start to finish. At least Edmund enjoys some small victories.

Lear, controlled by rather than in control of everything, especially (and most dramatically) the natural environment, loses his identity when he loses his ability to control spatial worth. Lear's dispossessing himself of his lands, his giving away of space, is a dispossession of masculine identity. This is very much in keeping with the ecofeminist idea that masculine identity often takes form in (and takes the form of) conquests of nature. As he loses his voice and identity, he becomes more unseated, more unhoused, and less distinguishable from the undomesticated spaces that wildly threaten civilization. Without his land, Lear becomes frenetic in his questions about his identity. In act 1, scene 4 alone, he asks three separate times about his identity in a crescendo of increasing frenziedness, first with a simple "Dost thou know me?" (1.26), then "Who am I?" (1.78), and finally, in desperation:

> Does any here know me? This is not Lear.
> Does Lear walk thus? speak thus? Where are his eyes?
> Either his notion weakens, his discernings
> Are lethargied—Ha! waking? 'Tis not so.
> Who is it that can tell me who I am? (ll.226–30)

The vehemence of nature's assaults hastens this old man's decline. This is the least of it, though. It is in the storm where we see him completely lose touch.

In the worst of the storm, the strongest example of an extremely obtrusive and hostile environment in the play, Lear sees homelessness as being the plight of other people in other places and not of himself where he is.[4] Though he tastes and smells the sulphurous air, hears the crack and spill of thunder, is blinded and burned by the lightning, drenched by the rains, and cooled by the winds;[5] though again and again and again, he is with the "houseless heads" (3.4.30) that are pummeled, buffeted, and pilloried by the hostile environment, yet he remains unable to see accurately. He is unable to see that home has become an impossibility for him. Lear is not in control, and the ability to control space is a prerequisite for home ownership. As Mary Douglas has argued, "home starts by bringing some space under control" ("The Idea of Home" 263). Lear does not have a clear notion at all about whom or where he is—his identity or the space he occupies. He is still pointing with third persons at other people: "Is man no more than this?" (3.4.102–3), he asks, pitying the miserable state of Tom.

Even his grand existentialism is a failure to perceive his own identity accurately: "unaccom-/ modated man is no more but such a poor, bare, fork'd / animal as thou art" (ll.106–8), he howls, tearing off his clothes as if to act the part that he already embodies and that he perceives as reality in Tom's "counterfeiting" (3.6.61). At Dover, unwilling to relinquish his voice, he blathers on and on in delusionary terms that he is what he was, claiming "I am the King himself /...Ay, every inch a king! /...I am a king" (4.6.83–84, 107, 199). If he is a king in anything but name, at this point, he is a sorry king indeed and is certainly not above the art he disparages: a coin would undoubtedly look better than the "side-piercing sight" (1.85) he presents. He is deluded about his identity. He is a mess, inside and out. He is a madman (SD.l.80), not a king.[6] If a house falls apart in a storm, we do not call the broken pieces a house: we call them broken pieces, debris, remains, or rubble. It can no longer be identified as a house: it has lost that identity. Lear, though he may still call himself a king, is a king in name only. Lear has divided the land that constitutes the core of his identity. The sad wretch who stares at the rubble and says "There's my house" is deluded. So is Lear deluded, staring at the rubble that was the king. Catherine Belsey explains that Lear's mistake has been in believing "that he can give away his kingdom and keep his kingship." Lear does not understand the relationship between land and identity and that "to name a person is to specify land, wealth, and the power that corresponds to them" (Belsey 54). Lear's space, home, power—his voice as an effective, functional king—are gone. Like Nixon's pondering of his fate without his bodyguards on the steps of the Lincoln Memorial in Oliver Stone's *Nixon*, the horror in *Lear* is one of displacement and disentitlement—specifically, displacement without an entourage into a world very far from home: the natural world. This horror is ecophobia dramatized.

The tragedies of the play depend vitally on the hostility of the natural world, and Lear loses his identities of kingship, male authority and privilege, and power to control what threatens him most insistently: his daughters and the horrors of the natural spaces into which they finally thrust him.

Horror is inseparable from and constitutive of the tragedies in *King Lear*. Quite apart from the graphic horror of blood and death in the play are the horrors of ontological unfixing and loss that nature poses. As Linda Woodbridge recently observed,

> In *Lear*'s England we lose our geographic bearings—for much of the play we do not know what kingdom we are in —and this radical de-centering after the opening in a recognizable center, the court, reproduces the loss of *social* place of those who become homeless.... Who we are is bound up in where Home is. Those who become homeless are strangers to themselves. (*Vagrancy* 296)

It is like looking into the mirror and seeing someone else's eyes instead of our own. Such is the horror of *Lear*, and it is resolutely environmental.

It is less that the play vilifies the kingdom willy-nilly, less that the play writes the land and the spaces that are so integral to British identities as arbitrarily horrific;[7] it is, rather, more that the play registers fears at the consequences of environmental unpredictability, resulting from radical crossings of boundaries. The horrors of "out there" are brought home, as it were, in the play. The socially self-consuming monstrosity implicit in the many metaphors of cannibalism in *Lear* reiterates the spatial and environmental dimensions of the developing tragedy. Cannibalism, always implicitly an environmental matter, voiced within an "out there" framework, is domestic in this play.

The play posits domestic disharmony both as monstrosity and as a form of cannibalism. The clearest articulation linking filial ingratitude, monstrosity, and cannibalism comes from the mouth of Albany, who maintains that if "these vild offenses" continue, "Humanity must perforce prey on itself, / Like monsters of the deep" (4.2.47, 49–50). Such, perhaps, is all well and fine in the vast expanses of the wilderness of seas and the rest of the natural world, but within the confined and carefully policed space of human society, it is a dangerous thing.

There are a lot of images of cannibalism in the play. We hear of a monstrosity that "to gorge his appetite" (1.1.118), as Lear complains, "makes his generation messes" (1.1.117). At another point, the Fool comments that "the hedge-sparrow fed the cuckoo so long, / That it had it head bit off by it young" (1.4.215–16). In addition, when Goneril displeases Lear, he tells Regan that her sister "hath tied / Sharp-tooth'd unkindness, like a vulture, [pointing to his heart] here" (2.4.134–35). Out on the heath, Lear tells Kent that " 'twas this flesh begot / Those pelican daughters" (3.4.74–75). The pelican, as the gloss in the 1997 *Riverside* edition explains, "was believed to feed upon its mother's blood" (1324). At another time, Lear commands that Goneril and Regan "digest the third" dower (1.1.128). At least at the level at which the metaphor works, Lear is guilty of commanding the very things that he has only just finished condemning, and the unnaturalness of the consumption he describes rankles our sense of justice. Certainly, nature is inextricable from all of this, and the discourse of cannibalism both reiterates and confirms the unwelcome intrusion of nature into the domestic spaces of the play, spaces which quickly dissolve and lose all traces of domesticity. Moreover, the self-consumption, logically, must result in a space of emptiness, nothingness—at best, an absence of culture and civilization; at its ecophobic worst, the horrors of an invasive and hostile nature.

If the discourse of cannibalism unsettles the spatial relationships between nature and home in disconcerting ways, Lear's calling Cordelia a stranger (1.1.115) is no less a spatial and environmental matter, one conceptually removing Cordelia from the space Lear identifies as home.[8] It goes against nature, in Lear's way of thinking, to have a child who is hostile to the domestic spaces he imagines, a child as obstinate, silent, and inexpressive as Cordelia, or as thankless as Goneril. The play as a whole seems to share Lear's view.

In a sense, Edmund, too, seems an unnatural thing, hostile to and divorced from the domestic spaces that the play imagines, but he is also decisively associated with an imagined nature that is essentially hostile, bad, and nonhuman. He expresses his allegiance (albeit, disingenuously) in terms of religious adoration: "Thou, Nature, art my goddess, to thy law / My services are bound" (1.2.1–2), he proclaims. Having been born outside of marriage, he falls outside the ideologically sanctioned space of home, outside of and banished from the privileges of property—and he wants back in: he is after his brother's land. He wants the space from which, at least in terms of inheritance, he has been banished. He calls on nature as the authority for his actions. In Lear's world, Edmund represents disorder, the chaos and horror of a world not in synch, where an unpredictable nature does not reflect and confirm human culture and society.

Through Edmund, we see the analogical thinking, of which a decreasing majority in early modern English society was so heavily enamored, intensely disputed. Whereas Gloucester maintains that the "late eclipses in the sun and moon / portend no good to us" (1.2.103–4), Edmund's thinking is of a different hue:

> This is the excellent foppery of the world,
> that when we are sick in fortune—often the surfeit of
> our own behaviour—we make guilty of our dis-
> asters the sun, the moon, and stars, as if we were
> villains on necessity, fools by heavenly compulsion,
> knaves, thieves, and treachers by spherical predomi-
> nance; drunkards, liars, and adulterers by an enforc'd
> obedience of planetary influence; and all that we
> are evil in, by a divine thrusting on. An admirable
> evasion of whoremaster man, to lay his goatish dispo-
> sition on the charge of a star! (1.2.118–28)

And yet, the fury of the elements that accompany the banished King implicitly maintains the validity of precisely the analogical thinking that so much else in the play unsettles. A King getting locked out by his daughters is not an everyday thing, and the storm that accompanies Lear on this strange night is equally unusual: "Since I was man," Kent exclaims,

> Such sheets of fire, such bursts of horrid thunder,
> Such groans of roaring wind and rain, I never
> Remember to have heard. (3.2.45–48)[9]

Even so, Lear cannot find "any cause in nature / that make these [Goneril's and Regan's] hard hearts" (3.5.77–78).[10] There is none. The causes are perhaps less in nature than in the time it took him to give his children their due. We will remember that Regan complains that Lear took his time about divvying up the inheritance (2.4.50).

Although both Lear and Edmund believe that nature patterns and reflects human behavior, Edmund challenges the terms through which this patterning is negotiated; Lear does not offer such challenges and, moreover, wants the analogical relationships to remain intact. Edmund's "excellent foppery" speech is a direct challenge, unequivocally verbalized, and it reflects the erosion of analogical thinking under the early modern winds of mechanistic change. Paul Delany's argument about "the struggle between the old order and the new" is useful here, and it explains that Edmund, Goneril, and Regan stand at one end of the spectrum, while Lear and his party stand at the other, and that this opposition conveys "a social meaning that derives from the contemporary historical situation as Shakespeare understood it" (34, 24). The triumph of the new is sheer horror to the old, and a large part of the horror resides in the new relationship between humanity and the natural environment, a relationship no longer of organicism but of competition and reciprocal conflict.

The two camps Delany speaks of entertain quite different notions, therefore, about the relationship between nature and domesticity. For Edmund, nature condones filial competition;[11] for Lear, nature condemns it. Whereas Edmund may pray to Nature, his goddess, whatever the gravity (or lack) of such an expression of devotion, it is what Peter Moore calls "an evil Nature" (171) that Edmund worships and feels he can, to some degree, control; for Lear, as both William Elton (126–27) and Eugenie R. Freed (51) have pointed out, when Lear calls on "Nature ... [to] convey sterility" (1.4.275–78) to Goneril, the sense is that this is not an evil nature, that it is one that exacts human justice. Moreover, unlike Edmund, for Lear, control is out of the question, and he perceives himself as a slave to nature, to fate, and to the winds of change. For Edmund, nature is serviceable; for Lear, its very unserviceability is part and parcel of the tragedy he suffers; but both characters clearly have their own ideological uses for nature.

All of the main characters in *King Lear*, to varying degrees, share a utilitarian view toward nature. It represents an object space that must be controlled, without which it is a dangerous space of chaotic nothingness. If Cordelia is associated with nature in the popular imagination that the play represents (or in Lear's imagination), it is certainly in this sense. Conceived of with the same ideals about silence as the natural environment (and valued analogously with it), women are for Lear a potently dangerous material, a space of poison and pollution that, like the natural environment, lacks reason, is morally inconsiderable, and must be kept silent. Cordelia is not silent.

For Lear (and, as we will see, for Albany), women and the environment are each viciously unpredictable and dangerous, and women who communicate freely are monsters.[12] No doubt, these imagined correspondences are a response to a heightened relationship imagined between women and weather at the time of *King Lear*, witnessed in the frenetic increase in witchcraft trials. Emily Oster has argued that "the most active period of witchcraft trials coincides with a period of lower than average temperature known to climatologists as 'the little ice age'" (216), and there is increasing archeological evidence that "the 1590s was the coldest decade of the sixteenth century" in England (Fagan 94). Weather was unpredictable and dangerous.

Two things (external to himself, at any rate) that threaten Lear most and ultimately bring him down are his daughters and the weather. "O, ho! Tis foul," Lear complains (3.2.23), that the weather "will with two pernicious daughters join / [their] high-engendered battles 'gainst a head / So old and white as this" (ll.21–23). The imagined monstrosity of insubordination of each (these women and this weather) before the king is more than simply a structural and thematic curiosity: it speaks to the question of agency and volition, which neither women nor the natural world can rightly lay any claim to in Lear's thinking.

Cordelia's silence, her artless "nothing," is something "which nor our nature nor our *place* can bear" (1.1.171—emphasis added). It is a subversion of authority for Lear that both nature and space repel, like the positive end of a magnet repels the negative. In Cordelia's "nothing," Lear hears something, and whatever her "nothing" signifies for him, whether her genitalia (as "nothing" signifies in *Hamlet*) or her status as a human subject,[13] the mere fact of her communicating *anything* is monstrous to Lear *because* she is a woman. Nowhere is his misogyny more clear than in what he reveals in his mad ravings to Edgar and the blinded Gloucester:

> Down from the waist they are Centaurs,
> Though women all above;
> But to the girdle do the gods inherit,
> Beneath is all the fiends': there's hell, there's darkness,
> There is the sulphurous pit, burning, scalding,
> Stench, consumption. (4.6.124–29)

Reminiscent of the "loathsome pit" of *Titus Andronicus* and all of its ambivalent (though mostly misogynist) significations, this "sulphurous pit," as Carol Thomas Neely has insightfully noted, is the accommodation place of the metaphorically confined fiend (*Distracted* 64). Unsanctified through marriage, it is "the dark and vicious place," Edgar explains, where his brother was begotten (5.3.173). At once an "indistinguished space" of nothingness[14] and a space very full of bad things, it is the source, presumably, of "woman's will" (of which Edgar speaks) and the "riotous appetite" (of which Lear screams).

In many ways, it is impossible to talk sensibly about the categories of "woman" this play constructs without an ecocritical lens. So much of what is going on with women in this play engages with notions about and ethics toward the natural environment. Virtually every failure of accommodation in this play operates through or is in some way associated with a construct of a full but empty, a silent but dangerously noisy, nature. Even the apparently well-housed Goneril and Regan, in fact, have accommodation problems. Cristina León Alfar argues that these sisters are figures of resistance who "are symptomatic of the patrilineal structure of power relations in which they live and to which they must accommodate themselves" (81), at liberty neither to dismantle nor escape the structures that house them.

The contorted logic defining nature and women operates on an ethics of anxiety about material predictability. Thus, control of the female body, Marion Wynne-Davies argues, was "paramount to determining a direct patrilineal descent, and when this exercise of power failed and women determined their own sexual appetites regardless of procreation, the social structure was threatened with collapse" (136).[15] The female body is the corridor in which wrestling, entangled discourses and contestations for power are played out, and since "the female body is the site of discourses that manage women...[discourses that are] continually working out sexual difference on and through the body" (Newman 4–5), then the mapping of the female body is obviously very important to our topic.

In a now old but still very relevant article, Peter Stallybrass argues that "woman's body could be both symbolic map of the 'civilized' and the dangerous terrain that had to be 'colonized'" ("Patriarchal" 133). The discursive dismemberment of women in the plays and sermons of the time[16] resemble not only the dismemberment and cutting of the land implied in maps and in the king's opening offer of land to his daughters; but it also resembles anatomies performed on animals and the dead objects of science. Yet, women remain, in many ways, unconfined, ultimately out of the reach of male control in this play,

and, therefore, within the ethical framework of the play, undomesticated, untamed, and unpredictable.

Understanding the imagined unpredictability of volitional or sexual women within a continuum of ecophobic images allows us to
see that the metaphors through which these associations are understood among scholars are themselves no less richly informed with
their own notions about nature, metaphors that reinforce the very
structures they seek to critique. Paul Delany, for instance, writes that
"though Lear has let the garden of England run to seed, it is clear
that Edmund, Regan, and Goneril have no interest in restoring it to
its proper condition" (34). The problem with this kind of metaphor is
that it reiterates the idea that the natural condition of nature (running
to seed) is not "proper," and it then proceeds to apply that notion of
impropriety and of a very human-mediated nature back onto the very
site of cultural reference on which its own identity depends.

The identities for nature and for women this play imagines are
accommodated in impossible paradoxes. The empty but full paradox characterizing the space of nature patterns the subjectification
of women in this play, but this is part of the larger patterning of
disaccommodation the play develops. To lose domestic space, to be
thrust into the natural world (conceptually or literally), to lose home
in this play means to be sentenced to exile from all of the rights and
privileges of human society into a hostile nature. Even the feigned
madness of Edgar shows this, perhaps more vividly, though, because
it is an over-the-top simulation.

The disguise Edgar has assumed, like so much else in this play,
is a compellingly ecocritical matter in which questions about power
are integral. As Tom, Edgar has taken "the basest and most poor-
est shape / That ever penury, in contempt of man, / Brought near
to beast" (2.3.7–9), and in his "nakedness" (1.11), he is as he perceives Bedlam beggars to be, as Lear himself, long before he tears
off his clothes (SD3.4.109), comes to be—unaccommodated, home-
less, banished from the community of humans into the wilderness
of the feared and unpredictable environment, into "the winds and
persecutions of the sky" (2.3.12)—and, essentially, without power.
The natural environment is a space of pollution that stands in stark
contrast to the clean and ordered environment of the community
of enfranchised people. It is a space in which Edgar will "grime
with filth" (1.9) his face, knot his hair, and stick himself with "pins,
wooden pricks, nails, sprigs of rosemary" (1.16) the better to blend
undifferentiated, a "horrible object" (1.17), into this space of ban-
ishment. He becomes a part of this environment, a thing devoid of

human identity, and complains "Edgar I nothing am" (1.21). The natural environment, so full of so many fearful things, is a space of nothing that disempowers and "make[s] nothing of" (3.1.9) those banished within it.

Edgar is able to begin shedding the assumed identity (of essential nothingness) that defined and shook him in the storm (though he still wears his disguise) only when he is no longer a pawn to circumstances that control him. It is roughly when he is in control, less tormented and more the master of his moves, that he is able to lead his father, though shakily, to the presumed cliff. Still, he does not have his identity, which, ultimately, is bound up with the control of his inheritance, the geographical space Edmund took from him: "Know," he tells the Herald, "my name is lost, / By treason's tooth bare-gnawn and canker-bit, / Yet I am noble" (5.3.121–23).

His father is less able to extract himself and regain the world he has lost. At Dover Cliff, Gloucester, deep in the guts of nature, is virtually consumed by his circumstances, and, like any consumed thing, lacks the power and autonomy of the consuming predator who can digest third dowers. Deep, deep in the guts of nature, he is totally without control, self-understanding, and identity. Pondering the great nothingness of death that waits beyond the edge of the imagined precipice, he is as far from home as he can be. Leading up to the cliff scene, Cornwall is acutely aware of the relationship between power/control and possession of home in his talk about "our power / ... which men / May blame, but not control" (3.7.25–27), but poor, naïve Gloucester still thinks Cornwall, Regan, and Goneril his guests: "You are my guests. Do me no foul play, friends. / ... I am your host" (3.7.31, 39). He never comes to understand his loss of space and the implications it has for his suffering.

Nor does Lear seem to understand very much by the end of the play. He is able to begin to resume, shakily, his identity, and with it new clothes, when more in control of himself physically, less at the mercy of the elements over which he has no control, the space and circumstances that completely control him. He is able then to talk without delusions about the uncertainty of his state and identity: "Would I were assur'd / Of my condition" (4.7.55–56), he laments. And the first questions he asks are about space: "Where have I been? Where am I?" (1.51). But he remains a pawn to circumstances and spaces and shows little change as a person. He knows Cordelia no better by the end of the play than in the first act; he merely knows his other two daughters better. He continues not to know Cordelia and to think that she hates him, though now he has revised his opinion about the cause of this hatred:

> I know you do not love me, for your sisters
> Have (as I do remember) done me wrong:
> You have some cause, they have not. (4.7.72–74)

He kneels and expresses his final bizarre fantasy of being with Cordelia in prison like lovers, not like a father and a daughter, and such is not what she offers: "We two alone will sing like birds i'th'cage, / ..." he says

> So we'll live,
> And pray, and sing, and tell old tales, and laugh
> At gilded butterflies, and hear poor rogues
> Talk of court news; and we'll talk with them too—
> Who loses and who wins; who's in, who's out—
> And take upon 's the mystery of things
> As if we were God's spies; and we'll wear out,
> In a wall'd prison, packs and sects of great ones,
> That ebb and flow by th' moon. (ll.9, 11–19)

It is his final fantasy, and it has everything to do with space, environment, and identity. He fantasizes about his place in the natural world where he is watching gilded butterflies, a safe little world full of sweet mysteries that he and his daughter will solve, but none of this bizarre fantasy matches up with his actual experiences of the natural world. The reality within the play is that the natural world is about as far from idyllic as we can get. It is a harsh world. Moreover, Lear fantasizes that he will finally have a daughter within a bounded space who will give him undivided love, of the sort about which Goneril and Regan spoke, and that he deserves such love. For all that he has been and will be punished; however, Lear has not changed the behaviors that brought on the punishment he has received. And for all the pity that he has shown, Lear is still a spiteful old man and does not "forget and forgive" (4.7.83), as he pleads for Cordelia to do; rather, spite-filled, bitter, and again deprived of his fantasy, he kills the person who hanged his youngest and, as Jonathan Dollimore correctly interprets (193), *boasts* about it (5.3.275).

From start to finish, the limits of Lear's identity and growth are staked out in spatial and environmental terms. It has been a struggle with boundaries, and we move from the very grand scale of nation and maps to the very personal scale of madness and an imagined prison. In between, nature triumphs and wipes Lear's slate clean of his eminently human pursuits: he loses power, identity, and home as much to unpredictable and uncontrollable nature as to his daughters and their ilk.

The horrifying specters of unhousing and alienation, the loss of identity and voice, and the almost apocalyptic chaos, all guaranteed by Cordelia's "nothing," evince a textual ecophobia as palpable as any of the characters on the stage or meteorological assaults on the heath. *King Lear* may well be "about power, property, and inheritance," an argument Dollimore makes (197), but it is no less so about an eminently unpredictable natural world, with bad weather, horrifying indifference, and countless dangers and paradoxes. The ecophobia at the core of our understanding and apprehension of the natural world is very central to everything that happens in *King Lear*.

3

CORIOLANUS AND ECOCRITICISM: A STUDY IN CONFLUENT THEORIZING

Coriolanus begins with a social and environmental crisis with which we have become all too familiar—famine—and dramatizes the subsequent revolts and ensuing impatience Caius Martius shows toward these. Critical responses to the opening scene have been varied, focusing heavily on the politics of the play's stance toward the hungry citizens; on the mimetic correspondences between the play and the 1607 Midland revolts to enclosure of grain fields; on the metaphor of the body politic and the language of corporeality in the play; and on the psychology of an isolated and besieged Coriolanus; but there has been little in the way of an ecocritical reading of the play's opening crisis and the subsequent crises to which it leads. This is surprising first because the confluence between matters of sexuality and environment are startling in this play and second because Coriolanus's isolation is deeply bound up with ecophobic issues.

PAIRINGS AND (ECOCRITICAL) IMPLICATIONS

Although much has been made of the last two syllables of the name "Coriolanus,"[1] whether we can sustain a case that the play's "scatological and anal discourse" (Paster 143) points unequivocally toward same-sex eroticism is less the matter in this chapter than is the environmental imagery and ethics involved in the scripting of bodies and contacts in the play.

Critics have, to be sure, offered queer readings of this play, but on no level does even the most provocative and novel of these—which would have to be Jonathan Goldberg's "The Anus in *Coriolanus*"—draw any kind of theoretical connection between the play's dual treatment of space and sexuality. This is perhaps not surprising, given the dearth of "queer ecocriticism" in general.[2]

On the surface, it hardly seems a matter of ecocritical interest that the only appreciable intimacy Coriolanus—anus of the Corioles—develops is with Aufidius and that it is, to some degree, mutual in its inclusions, exclusions, and the kind of space it creates. The love that Aufidius speaks of for Coriolanus, his imagery of twining bodies and hot contestations points to a physical and emotional relationship in and with which women not only have no part or place but also clearly cannot compete. "Know thou first," Aufidius explains, deftly writing women out of the picture,

> I lov'd the maid I married; never man
> Sigh'd truer breath; but that I see thee here,
> Thou noble thing, more dances my rapt heart
> Than when I first my wedded mistress saw
> Bestrid my threshold. (4.5.113—18)

Coriolanus feels much the same way and, approaching Aufidius's house, comments on the "slippery turns" of the world:

> Friends now fast sworn,
> Whose double bosoms seems to wear one heart,
> Whose hours, whose bed, whose meal and exercise
> Are still together, who twin, as 'twere, in love
> Unseparable, shall within this hour,
> On a dissension of a doit, break out
> To bitterest enmity; so, fellest foes,
> Whose passions and whose plots have broke their sleep
> To take the one the other, by some chance,
> Some trick not worth an egg, shall grow dear friends
> And interjoin their issues. (4.4.11—22)

Presumably, such a switch from enmity to "love" (1.23) envisions the double bosoms of one heart, the hours together, the bed, the twinning, and the inseparability of friends fast sworn.

Robin Headlam Wells makes a case against a queer reading here, arguing that "since Martius and Cominius are characters in a play, not real people with a life beyond the stage, it makes more sense to enquire what their metaphors have to do with war, military values, and social conflict than to ask how these men spend their hypothetical private moments," and that "although it is not at all uncommon to find knights in medieval romance embracing and kissing with all the apparent fervor of lovers, such extravagant displays of affection are not normally a sign of sexual interest" (411).

Yet, we need to remember, as Lisa Lowe reminds us, that this is a play "about the impossibility of achieving an absolute, singular manhood" (89); a play that radically questions both gender and sexual identity—indeed, the former through the latter, resulting in "emasculation and submission" (Lowe 94) for Coriolanus. If gender stability were not so front and center an issue in this play, Wells would surely have a viable point, but the fact that identity is such a slippery thing here makes the queer moments much more important than Wells seems willing to accept. These moments, situated as they are within a discourse irreducibly environmental, require ecocritical unearthing.

If the play makes a "rhetorical comparison between the body politic and the organic body [that] naturalizes the inequality of power between patriarchs and plebians," as Lisa Lowe claims (91), no less does it extend a naturalized ontology of differential power relations to sexual alliances. In the case of Aufidius and Coriolanus, the relationship is obviously physical, obviously same-sex, and obviously deeply involved with matters of space, geography, and environment— the objects of the enmity between these two men. At the same time that the relationship between these two men is determined by their relationship to the physical spaces that surround them, the space of their love itself (however we define that love) is a dangerous one, an uninhabitable space, a space neither of heterosexual marriage nor of same-sex friendship, a contemptible space somewhere between Rome and Corioles, a space that cannot be voiced without revulsion. The play generates precisely such a revulsion toward Coriolanus, eventually resulting in his expulsion. The contempt the play generates for its eponymous hero, though, is obviously not the contempt of homophobia in the twenty-first century sense but the seventeenth-century contempt of a sexuality that is linked with inconsistency and slippery turns.

Part of this inconsistency no doubt has to do with what Alan Bray calls "the immense disparity in this society between what people said—and apparently believed—about homosexuality and what in truth they did" (9). Part of it may also have to do with Elizabethan notions connecting homosexuality[3] with disorder, with earthly experiences, with wildness, and with no small degree of unpredictability.

Coriolanus is the face of this disorder and unpredictability. In the end, he becomes a thing gone wrong, an ugly blemish on the body of nature, "a disease that must be cut away" (3.1.294), a foot "gangren'd" (1.305), an "infection...[that might] spread further" (ll.308–9), a weed-like aberration that must be plucked (1.307). Indistinguishable from the natural world, he becomes a beast that angry (and hungry) opponents may "tear...to pieces" (5.6.120). And to be associated with such a world is no happy thing.

Nature abhors the queer in this play. Coriolanus seems unnatural because he is unnatural: "Tell me not wherein I seem unnatural" (5.3.83–84), he commands because he knows. He is at home nowhere. And he will listen to no one. Again and again and again, he denies others their voice. He is a thing alone, "made by some other deity than Nature" (4.6.91), according to Cominius, but a deity "that shapes man better [than Nature does]" (1.92). The play, however, tells us otherwise. Coriolanus has no power over and is no better than nature. He can talk about inversions, pebbles on hungry beaches filloping stars, winds striking cedars against the fiery sun, and so on, but the reality is that this boy of tears is an infection that must be "cut off" (5.6.138), like Dr. Faustus, the branch that might have grown full straight, as it were. It is perhaps another echo of *Dr. Faustus* that all of the people around Coriolanus at the end of the play are screaming "tear him to pieces" (1.120). We will remember that the good Doctor Faustus, an egomaniac who wants the power of a god, has his limbs all torn asunder by the hand of death. Coriolanus is expendable precisely because his individuality,[4] accentuated by his suspect and certainly minority sexual position, puts him in direct conflict with the community that expects his leadership. He really is a weed that starves his ecosystem.

In a startlingly provocative paper presented in March 1999 (see Works Cited), Danne Polk asks questions about the symbolism of weeds:

> It is important to understand how this symbolism works. What is a weed? How do weeds come to be identified in nature? How is it that certain material bodies come to be defined as unwelcome even though they exist? How do sexual minorities, even though we exist, come to be defined as a contamination of nature? As "against nature"? (http://www.queertheory.com/theories/science/deconstructing_origins.htm)

"In nature," as Emily Compost succinctly puts it, "there is no such thing as a weed" (http://www.emilycompost.com/weed_definition. htm). The concept of "weed" is a socially useful construct that writes nature in terms of its utility to people. Common dictionary meanings have weeds as things that have no practical value to people (i.e., they do not produce edible materials or attractive adornments or otherwise commodifiable products). Moreover, weeds often express a parasitical relationship to other plants or their food, making the "weed" stronger and killing the other plants or making them weaker. Though Coriolanus is never explicitly called a weed, effectively and symbolically, he functions as one. He has no place, no home, in the world of the plebeians.

If Elizabeth's statutes against sedition[5] are any gauge of what was going on in the Tudor body politic, the plebeians of *Coriolanus* represent the threat of organized action from the people at large against the state, a threat that for Coriolanus is not merely political but, more importantly within the dramatic action of the play, a threat to his sense of home. It is a threat that is carried out.

The topical relevance of the plebeians' potential for organized revolt running as a subtext in their plea for corn voiced an anxiety with which contemporary viewers were amply familiar. Numerous critics have noted that the revolt for food in the play had to have resonated with the audience because of the 1607/8 uprisings in the Midlands, uprisings whose immediate goal was, according to E. C. Pettet, "the breaking down of enclosures" (34), and whose motivations were, perhaps as F. Gay suggests, related to issues of dearth.[6] From the start, the dramatic action of the play is environmentally motivated. Finishing with Coriolanus as a queer has several effects.

If we agree that Coriolanus functions symbolically by the end of the play as a weed, then his erotic position must also put him in "the category of disposable commodity, essentialized as invasive threats to the purity of the system itself" (http://www.queertheory. com/theories/science/deconstructing_origins.htm). Certainly he seems a thing of the natural world, an object exploited and then disposed of when its utility has expired, an object accorded the same moral status of the natural world. He is a leader who becomes a dismembered carcass, having been consumed by his people. Having been consumed, he is, perhaps, the "disposable excess" Jonathan Goldberg speaks about ("The Anus" 262) by the end of the play. Moreover, the way Coriolanus is gendered complicates, as Goldberg also contends, "the division natural/unnatural, with female and male as 'mothers'" ("The Anus" 268). The argument Goldberg makes is that Coriolanus is "doubly gendered" (ibid.), and there is a lot of evidence to support such a claim. He does, after all, have to prove himself a man (1.3.17) to his mother; he has to get beyond his virtual "Amazonian" status (2.2.91) and *not* "act the woman" (1.96); he is feminized by the repeated masculine penetrations against his often bleeding body;[7] he is doubly gendered. Goldberg argues that

although orality has figured so heavily in discussions of the play,
anality

> is the *productive* site for the character in the play and not merely
> to be read within the Freudian trajectories of shame and sublima-
> tion. Coriolanus's career of attempted self-authorship represents a
> desire to become a machine, to "live" in some realm that is not the
> biological. (268)

Figured as a site of production, Coriolanus is indeed a confusion of
natural and unnatural, bowels and wombs, bodies and machines,
and so on. As such, he is evacuated from the body politic. Again,
nature abhors the queer in this play and seems in some ways to invalidate sexual minority positions while validating majority positions.
There is a confluence here in the scripting and valuing of sexualities and of nature itself. Calling on "nature" to validate heterosexual
mores "would," as Greta Gaard argues, "seem to imply that nature is
valued—but . . . this is not the case" ("Toward a Queer" 120). Taking
Gaard's argument a bit further, we can see that rather than authoriz-
ing and valuating nature as something above the realm of the human,
something determining of social values, something to be respected,
validating sexuality through nature is a devaluation of the natural
and a failure to see nature on its own value-free terms. Failing to see
how nature is constructed in these discourses of sodomy, how nature
is determined *by* social values, and how speaking for nature and making it speak for Henry or Elizabeth or Bowers or Georgia and against
Coriolanus, places nature in the discursive position of legal tool and
commodity in the arsenal of things men use to make and enforce
their laws.

Any kind of commodification of nature must be inherently ecophobic, in the same way that any commodification of women must be
misogynistic and of sexual minorities, homophobic. What is amazing
is the number of connections between discourses that write the natural world in this play and those that write sexual minorities, and it is
more amazing that there has been no critical work until now that has
analyzed these connections.

Although clearly there is more work to be done, it is also clear that
the environmental ethics the play reifies and the crises to which those
ethics ultimately lead—the drama's discursive playing of nature as
a space of weeds out of which Coriolanus is plucked and undone—
valorizes imagined notions both about what good nature and good
sex are, neither of which characterize anything about Coriolanus.

On a very basic level, one easily understood in our undergraduate classrooms, it is clear that the play constructs a space that cannot tolerate either the fierce individualism that so tragically characterizes Coriolanus or the position of sexual minority into which he finally thinks he will have the consolation of escaping. The writing of such spaces in such ways is at once a rehearsal of ecophobia and a scripting of normative sexualities.

ECO/EGO CONFLICT

Michael Bristol has noted of *Coriolanus* that "the play is saturated with concrete situations in which the fate and condition of bodies is of paramount importance" (213). But importance to what? In Bristol's view, the various body images the play presents are central to an understanding of the play's ideological stance toward the hungry plebeians. He argues that "the ambient condition of political uncertainty and of gathering resistance to lawfully constituted authority is expressed in a complex and ambivalent rhetoric based on the traditional *topos* of the body politic" (ibid.). Zvi Jagendorf also observes that the play obsesses on the body, that "everywhere we encounter legs, arms, tongues, scabs, scratches, wounds, mouths, teeth, voices, bellies, and toes together with such actions as eating, vomiting, starving, beating, scratching, wrestling, piercing, and undressing" (458). Jagendorf argues against Michael Bristol's pleb-centered reading of the language of the body in the play and claims that the significance of the body trope is "to generalize the state of social fragmentation enacted in a specific case on the stage and to isolate the lonely figure whose withheld body is the subject of so much attention" (458–59n5).[8] Certainly, Coriolanus *is* that lonely, isolated figure, fragmented from, a fragment of, a larger, fragmented social body; moreover, it is impossible to talk about Coriolanus's alienation outside the context through which such alienation takes its form and meaning. But the play does not begin with Coriolanus; it begins with the fragmented social body, and so to argue as Jagendorf does that the play generalizes *from* Coriolanus *to* the social body is inaccurate, and we could more easily and accurately make the exact opposite argument and claim that the body trope generalizes the state of one man's isolation enacted in a specific case of a fragmented society. Still, if the plebeians are "fragments" (1.1.222) in Coriolanus's thinking, Coriolanus himself is no less a fragment in the thinking of the play, and he is as much the effect of forces beyond his control, is "no surer" (1.171), to use his words, than the plebs he condemns. To read him as a case study is to take sides with the individualism he champions. The play, rather, presents him as the effect of both nurture *and* nature, and it is the struggle between the two to produce him that initiates and sustains the drama.

And if there is one thing that the play teaches us, it is that the production of the subject is also the production of the natural world, and I would like to reiterate that herein lays one of the fundamental differences between ecocriticism and Nature Studies. Whereas Nature Studies takes the natural world as a primary object of study (to the point that cries can be heard for having nature an actual protagonist in literature), for ecocriticism, in contrast, nature is always part-and-parcel with the writing of subjectivities. Moreover, since nature is always conceptualized *through* people, through subjects, is always the product of specific social and historical configurations, is always, in a word, *constructed*, it is both meaningless and inappropriate to pretend that nature exists in some sort of prediscursive realm in texts that *we*, as subjects, produce. This does not mean that subjects are front-and-center in the discussion or that culture exists *a priori* to nature; rather, the social is *embedded* in the natural. Jean Arnold explains persuasively that "To disregard the fact that human cultural production is embedded in the natural world is to entertain a selective vision that places humankind in a pre-Copernican position of centrality it does not deserve" (1090). It is virtually impossible to separate the writing of subjectivities from the writing of nature. And whether or not there is an explicit reference to Nature in a given text, whether or not nature has a voice, it is there. Part of the task here is to determine what the "it" is that is there in particular texts, what kind of nature is being produced, how it fits into the moments of history that produced it, and why that nature and those moments of history are relevant to the present.

The imbalances and fragmentations in the life of Coriolanus write the natural world as a quirky, fickle kind of a place where we can find radical inconsistencies, where "a lamb indeed...baes like a bear" (2.1.11), where crows sometimes peck eagles (3.1.139), and "an unnatural dam" will occasionally eat its young (3.1.291–92). It is pure analogical thinking that this matter speaks to; we see here the ideology of the text in conflict with that of its eponymous hero. Coriolanus seeks a firm division between the famous antagonists, nature and nurture. Consistently, the play urges otherwise—namely, that Coriolanus is an agonistic effect of both *essence* and *ideology*, and that Coriolanus's failure to strike a balance between the two is *the* sole cause of his problems.

The degree to which Coriolanus can play the parts he has learned is limited by his ability to remember those parts. At times, he forgets them.[9] He may want to separate himself from "the beastly plebeians" (2.1.96) and the natural world that they signify, but we just do not get the victory of one side or the other. He says he won't "be such a gosling to obey instinct" (5.3.35), to be, in other words, the effect of nature, yet, he confirms nature as an authority whenever he uses the term "unnatural" to describe social affairs (5.3.84, 5.3.184). Even though Jonathan Dollimore could hardly be more right in claiming that "essentialist egotism, far from being merely a subjective delusion, operates in this play as the ideological underpinning of class antagonism" (222), an egotism that is also "a complex function of social relations" (ibid.), it is very much against the text to claim, as Dollimore also does, that Coriolanus "is in fact the ideological effect of powers antecedent to and independent of him" (218). Such a claim insists on the victory of the constructionist side of the constructionist/essence binary, but it is precisely the absence of such a victory that enables the deep and crippling fissures in our hero. Fragmented from society, unhoused, and stripped of an authoritative voice, Coriolanus is the displaced effect of competing discourses about what "home" means in this play.

HOME AND "A WORLD ELSEWHERE"

To some degree, home is a place whose sole purpose seems to be to nurture men in the early modern period and in *Coriolanus*. Virgilia seems to exemplify this model. She is a perfect example of the good wife that early modern conduct books typically described. As Georgianna Ziegler explains,

> Women were not only enjoined to remain in the house, but when they had free time, they were to spend it in their chamber, not given to idle occupations (which male pamphleteers were sure would lead to vain imaginings), but engaged in prayer or useful household pursuits, such as sewing...The good housewife was literally the *house keeper*. (77, 76)

Of course, Virgilia won't go out and play, and Valeria's earlier comment that Virgilia is a "manifest house-keeper" (1.3.51–52) is an apt description. Virgilia, who is, as Janet Adelman remarks, of "relative unimportance" (324n46) in the play, is uncharacteristically outspoken to Valeria and Volumnia about her will to stay indoors: "I'll not," she says unwaveringly, "over the threshold till my lord return from the wars" (1.3.74–75). Closed off from the world, she is one of the commodities of the controlled space that home (and prison, for that matter) represents. Home in this play, as in the conduct books, is a safe place (for men) where men are in control and women are in submission; "outside" and nature are the domains of men; "inside" and domesticity are the domains of women (under the direction of men, of course). Although Coriolanus does not endure the domestic incarceration and lack of connectedness with the natural world that Virgilia does, home is a dangerous place for him, perhaps the more so because it is so clearly not a place for him. His desire to have a home is a kind of reversal of roles that can only result in tragedy.

If home is, on the one hand, a place of nurturance for men in this play, on the other hand, it is a profoundly dangerous place for Coriolanus when he tries to cross the boundary that separates the spaces the play assigns men and women. Home in *Coriolanus* is more dangerous in some ways than the natural world because, like the Rome of *Titus Andronicus*, it houses a domesticated wilderness, a dangerous place where everyone is hungry, and "everyone seems in danger of being eaten" (Adelman 154).[10] Such is certainly the way of the natural world, but it is not what we expect in our homes: "home" walls off the perceived economy of a natural ecosystem, where everything is fair game. Home is not a place where people "feed on one another" (1.1.188), and Coriolanus is both aware and afraid of the metaphoric cannibalism he senses in Rome. Everything living in the natural world sooner or later becomes food, but "nature teaches beasts to know their friends" (2.1.6), and home is where friends are.

Although the concepts of ecology and of the flow of energy in ecological systems postdate Shakespeare, there is wide, if rudimentary, commonsense understanding of such energy maps, of "the crows...[that] peck the eagles" (3.1.139), and so on. Terrified of this world, this "city of kites and crows" (4.5.42), Coriolanus flees. And he has good cause to be terrified of this home, this Rome.

There is something a bit spooky about Volumnia and her bizarre feeding habits, her supping on herself, not to mention on her son. There is something unsettling about her amoral bloodthirstiness. Her description of Hector could not be more vivid:

> The breasts of Hecuba,
> When she did suckle Hector, look'd not lovelier
> Than Hector's forehead when it spit forth blood
> At Grecian sword, contemning. (1.3.40–43)

There is something not quite right about the pleasure she, Valeria, and Menenius take in the "hurts i'th' body" (2.1.150) of Coriolanus.[11] There is something wholly discomforting about the warmth of her patriotism and how it cools her maternal concerns about the physical pain and mortal danger Coriolanus has to endure. There is something unseemly in a parent who clucks her son to the wars (5.3.163) and then complains to him "thou has never in thy life / Show'd thy dear mother any courtesy" (ll.160–61), insisting that "there's no man in the world / More bound to's mother" (ll.158–59) and trying (with Valeria and Virgilia) to "shame him with our knees" (l.169). Coriolanus knows the dangers of his "home"; he also knows that "there is a world elsewhere" (3.3.135), that he must not stay at home (although he craves home desperately), that he has to go away and search for what he needs. It is an archetypal male pattern in literature, where the man has the horizon to work out his crises.

Coriolanus clearly needs a different kind of home, a different kind of mother, and a different kind of past. In an analysis of relationships among race/culture, environmental exploitation, and colonialism, Maria Mies draws an interesting connection between what she calls "nostalgia for childhood" and the search for "home": "The nostalgia for childhood and the search for motherliness," she argues, "are often combined with the search for *homeland* or *home*, for belonging, for one's own place" (141). The metaphor equating childhood with ideals of goodness and innocence is a useful one to apply to Coriolanus. Given what we know of him—his relationship with his mother, his emotional immaturity and infantile egotism, his evident lack of a warm and nurtured childhood,[12] and his lack of a satisfactory place to call "home"—we can understand his self-banishment as a search, an exploration into geographical spaces that function as his resource and are available for him to exploit. Written into his knowledge of "a world elsewhere" is an ethics of domination, a colonialist imperative whose basis resides in the egotism of which Coriolanus is so well-endowed.

He is essentially selfish and concerned only with what his people can do for him. Divided in his loyalties, severed from any sense of place that he may call home, and unable to effect the self-fashioning his fierce individualism craves in any way that his community can tolerate,[13] Coriolanus is unable to integrate, to show or accept pity, or to offer or listen to speech. This spells his undoing.

ALL UNDONE

Voicing a running motif of the play, Menenius fears that he and everyone "are all undone" (4.6.107) unless Coriolanus shows mercy.

But it is very clear that more than mercy is at issue here: it is a thorough inability to integrate that *Coriolanus* thematizes. From the fragments of act 1, to the fears of undoing, and finally to the undoing of Coriolanus, the threat to "tear him to pieces," and the killing of him, this is a play about disintegration into the imagined belly of a terrifying nature. Questions about relationships between speech and silence, voicing and listening, and reason and feeling: these are clearly pivotal in the undoing of the eponymous tragic hero of the play—and as with so much else in this play, matters of environment here are central.

At the close of *King Lear* and its tour-de-force of environmental hostility and assaults against poor passive humanity, Edgar's primary solution to the silence that motivates all of the dramatic action is that we should "speak what we feel, not what we ought to say" (5.3.325); in *Coriolanus*, we see what happens when people *do* live by such unrestrained egotism as Edgar endorses. Coriolanus speaks "in bolted language; meal and bran together / He throws without distinction" (3.1.320–21), "with a voice as free" (1.73) as he wants. In an age that champions precisely such fierce individualism, Coriolanus is the triumph of the ego. The only problem is that ego and environment are totally incompatible in this play. For ecocriticism, the triumph of the early modern ego is the triumph of violence.

It is not just the violence toward controlling his own body that this play reveals but also toward the world from which that body comes and on which it depends. Control over nature defines a key shift in the relationship between people and the environment in the early modern period. "The image of nature that became important in the early modern period," Carolyn Merchant observes, "was that of a disorderly and chaotic realm to be subdued and controlled" (127), and with the advent of science, it became mandatory to extend control, deepen exploitation, and normalize hierarchies. Essential to accomplishing this task was to intensify anxieties about an uncontrolled nature, about the monstrous results of letting nature have its free and riotous reign. These antinature anxieties are integral to the rigid boundaries developing in the early modern period. They both result from and fuel an ethics toward the environment that is heavily ecophobic—which is not, of course, to say that it is only ecophobia that defines early modern relations with the natural world. Yet, while there is wonder—a lot of wonder—and joy that are vividly and vigorously expressed in Shakespeare, in pamphlets, in poems, in song, and in many other early modern sources (much of which has been well-discussed), I am concerned less with the biophilic or the ecophilic impulses that obtain in Shakespeare than with the various forms of expression the plays generate of ecophobia. Again, however, it is necessary to reiterate here that ecophobia in these discussions is not intended as a monolithic paradigm defining necessarily the primary way that humanity responds to nature. Rather, the intent here is to show convincingly that among the many changes occurring during the early modern period, one was a clear and measurable increase in ecophobia and that Coriolanus is symptomatic of this early modern world and the changes that were happening in it.

Symptomatic of the time in which it was written, *Coriolanus* offers a hero who draws a separation between himself and everything else that is very sharp indeed. The irony, of course, is that this separation Coriolanus seeks between self and world backfires. Coriolanus esteems himself too highly, is, as his mother says, "too absolute" (3.2.39), as if he "were a god,[14] to punish; not / A man of their [the plebeians'] infirmity" (3.1.80–82). For Coriolanus, the natural world is not a pretty place: it is an ugly and constant reminder of weakness, and it is filled with plebeians. There is nothing worse than plebeians in his thinking. They are the very worst with which a dangerous natural world threatens safety, domesticity, order, and, of course, individuality. For him, the plebeians are "souls of geese, / That bear the shapes of men" (1.4.34–35) and are indistinguishable from a demonized environment that looms waiting to reclaim geographical and ontological spaces, to reformat the program Coriolanus so acutely seeks to maintain: "You common cry of curs," he hollers,

> whose breath I hate
> As reek a'th'rotten fens, whose loves I prize
> As the dead carcasses of unburied men
> That do corrupt my air. (3.3.120–23)

He calls them "those measles / Which we disdain should tetter us" (3.1.78–79), and curses them that

> all the contagion of the south light on you
> You shames of Rome! You herd of—Biles and plagues
> Plaster you o'er, that you may be abhorr'd
> Farther than seen, and one infect another
> Against the wind a mile. (1.4.30–34)

Common, bestial, rotten, and contagious, they are the noise to which he imagines he must voice himself. They are the very disorder of nature, the imagined chaos that lies outside of human society and culture, a vast interconnected network of cacophonous and common voices against which his lonely individual voice seeks its own kind of order. Speaking as one, they are the imagined undistinguished space of the natural world; yet, for all of his contempt, there is little in the play to validate his ecophobic scorn; rather, such fits of name-calling suggest both an awareness of and insecurity about his own embodiedness, fragmentation, and isolation, and it forces a focus on his body. It is this that becomes the site of confusion of natural and unnatural, a commodity unfit for both social and natural economies, and thus disposable. It is his body that is disciplined and tortured, manicured and maimed, the object of this play's ecophobic fury. It is his body that finally functions as both his own voice and the voice of others.

Notwithstanding his monstrous ego, Coriolanus is not very good either at speaking of himself or hearing others speak of him—or of anything. He has very serious issues with speech and himself admits that "when blows have made me stay, I fled from words" (2.2.72). He *is* the "boy of tears" (5.6.100) Aufidius calls him, a narcissistic and smug little brat who is singularly incapable of hearing himself criticized, being named, or being spoken about. It is less what Janet Adelman calls "oddly touching" (156) than it is pathetically reveal-ing when Coriolanus, insecure and needing to know what Aufidius thinks of him, asks "Spoke he of me?" (3.1.12).

Though he desperately needs to know what people say of him, Coriolanus simply cannot listen. Listening is fatal to him. It undoes his sense of self. He orders Menenius quiet as his internal conflict rises to crisis: "Another word, Menenius," he says menacingly, "I will not hear thee speak" (5.2.91–92). In his alliance with Aufidius, he tries to stop his ears, claiming that "nor from state nor private friends, hereafter / will lend ear to" (5.3.18–19). He is unable to listen to the angry plebs and would (and does) sooner accept exile than relinquish the control that having deaf ears affords him. And when he does lis-ten, it spells his defeat: as he tells his mother,

> You have won a happy victory to Rome;
> But for your son, believe it—O, believe it—
> Most dangerously have you with him prevail'd
> If not most mortal to him. (5.3.186–89)

He is, of course, right, and, having conceded to his mother's pleas for him to listen, he dies a few scenes later.

Petitions to be heard begin the play, with a Roman citizen urging his fellow citizens[15] to hear him speak (1.1.1–2)—and they do. He speaks freely, "not maliciously" (l.35), "in hunger for bread, not in thirst / for revenge" (ll.24–25). We move very quickly from hearing an individual pleading for the right to speak and be heard to hearing the general plebeian body making such a plea. Menenius tellingly responds by asking "Will you undo yourselves?" (l.63). Within the Roman power structure that the play imagines, speech is a potentially dangerous thing that can result in an undoing of the safety that the social fabric promises, the very structures that constitute "home" for the plebeians.

The surgical divisions in the land and society—an implicit and running theme in this play—not only remain but are exacerbated by the end. It is perhaps not wrong to claim, as Arthur Riss does, that "Coriolanus falls because he asserts himself as a private, absolutely enclosed, literal 'body' in a society that mandates he embrace an ideology of the body politic" (54). In so doing, Coriolanus is fragmented from—a fragment of—a larger, fragmented social body and seeks "a world elsewhere" (3.3.135), but there is nowhere for him to go. Devoiced and unable to keep his footing in the slippery world, he is victim of what he champions most. Valor is the unnatural dam that eats up this child who "might have been enough the man.../ With striving less to be so" (3.2.19—20), this boy, this not-quite man. Yet, it is his very embeddedness in a disruptively undomesticated, inhospitable, and, in many ways, uninhabitable environment that posits a crisis.

Though less obvious at a glance than in *King Lear*, home and space are key players that are deeply involved with the matters of voice and environmental ethics that this play raises. Home manages human/ nature relations in this play, but nature itself is a construct, a viciously punitive source of authority. It is the space and origin of authority against which Coriolanus cannot be "author of himself" (5.3.36); a space of weeds, out of which Coriolanus is plucked and undone; a space that cannot tolerate either the fierce individualism that so tragically characterizes Coriolanus or the position of sexual minority into which he finally thinks he will have the consolation of escaping. In the end, there is no escape: nature wins, and it is a very dangerous and consuming nature.

4

PUSHING THE LIMITS OF
ECOCRITICISM: ENVIRONMENT AND
SOCIAL RESISTANCE IN *2 HENRY VI*
AND *2 HENRY IV*

I wonder what we are doing—Shakespeareans and non-Shakespeareans alike.

—O'Dair *Class* 101

At the 2009 ASLE conference in Victoria, BC, at the first of the two "Soul Food" sessions chaired by Sharon O'Dair, after an increasingly heated Q&A period, I asked the speaker—Andrew Battista—in exasperation, "well, if you're *not* here because you want to *change* things, then why *are* you here?" His immediate reply seemed to astonish everyone in the room: "I'm *here* 'cause I want a job. I do ecocriticism because there's a *market* for it." It had long been a fear of mine that things would come to this. In a debate with Leo Marx at the 2003 ASLE in Boston, Lawrence Buell said, "I'm sure there's no one here" for professional advancement—without even a hint of self-parody or insincerity—and I thought, "You *must* be kidding?" He has, since then, modified his comment somewhat, suggesting that professionalism may, in fact, be a reason why people do ecocriticism, but that there has to be more to it than that "criticism worthy of its name," he explains, "arises from commitments deeper than professionalism" (*FEC* 97).

When Battista went on to defend his position, of course, I had to admit that it is a good thing for ecocriticism to have achieved enough scholarly acceptability that Shakespeareans now embrace rather than reject out-of-hand green activist approaches; it is a good thing for ecocriticism to have achieved enough attention that ecocritics will actually be able to find employment; and it is a good thing to have more people doing more and more diversified work. Yet, at the same time, there are the Battistas, and O'Dair is not alone in her worries: fellow Shakespearean Annabel Patterson asks, "To what end" (10) we do what we do, and it is difficult to avoid the feelings of "ecodespair" Scott Slovic mentions in his Foreword to *The Greening of Literary Scholarship*.

If ecocriticism is to be more than professional exhibition-ism and intellectual masturbation, then it must connect with the world outside of the text in some meaningful way. Such a con-nection is no doubt what fuels the thick description approach of New Historicism. Uncovering ties among structures of oppression is, perhaps, the crux of resistance and activism for academics. Yet, even as we acknowledge this, it is difficult to know how much or what kind of environmental impact literary scholarship can possibly have, especially of Shakespeare. In an article entitled "Ecocriticism as Praxis," David Mazel, apparently doubting the relationship that learning and knowledge has with activism, asks for evidence from "empirical research" to prove that "students who read and write about green texts turn into more thoughtful and effective environ-mentalists than they might have been otherwise" (42). He is dis-turbed at not finding such evidence. The problem with this kind of logic, however, is that it frames the issue within an absolute binary wherein measurability becomes the sole source of knowledge-making, and this is surely invalid. I can say personally that I do not know anyone who has done anything measurable as a result of my ecocritical teaching or writing; by the same token, though, I do not know anyone who has committed murder because of watching violent action movies.

Devoted to an activist stance, ecocriticism has gone far, but it can go further, and we need to continue to push its limits. Far, far from the familiar terrain of nature writing and Edward Abbey and Gary Snyder and Leslie Marmon Silko and other such favorites of ecocritical scholarship, we have Shakespeare's *2 Henry VI* and *2 Henry IV*. These plays, which draw on what Patterson has called "a cultural tradition of popular protest" (38),[1] offer ecocriticism a good chance to push its analytical limits, to look at how and why class struggle and environmentalism can profit from working together, and to discuss the thematically linked matters of disease, social resistance, and environment on theoretically equal terms, rather than to pit coal miners against spotted owls (as O'Dair does) or vice versa or to argue for which is more "deserving of our atten-tion" (*Class* 108).

PART 1: *2 HENRY VI*—TROUBLE IN THE GARDEN OF IDEN

Social protest in early modern literature often confirms the values of the court/country binary, where the court is a space of back-biting, corruption, and competition, while the country is one of respite, purity, innocence, and so on. In *2 Henry VI*, however, as Thomas Cartelli has suggested, "the notion of the garden . . . as an unturmoiled place apart, untouched by the social strife that reigns elsewhere [in the play], . . . becomes radically qualified" (52). We get less of the uncontested "familiar contrast of court and country" that Stephen Greenblatt speaks of ("Murdering" 124) than "a space intersected by mutually exclusive and competing class interests" (Cartelli 52).

The trouble for Jack Cade begins long before the garden of Iden, and long before we actually see Cade. Virtually every scene in the play is in some way a working out of some kind of rebellion. Rebellion is what the play is *about*, and in a sense there is very little distinction among the forms that rebellion takes in the kind of world *2 Henry VI* presents. Paola Pugliatti is surely correct in asserting that "the play suggests a leveling to the lowest plane of those who intrigue at court for their own advancement and profit and of those who rebel out of material need and hunger" (456). Certainly the metaphors and settings for rebellion in this play substantiate such a claim. Queen Margaret describes Gloucester and the "commons' hearts" (3.1.28) he has won as shallow-rooted weeds: "Suffer them now," she warns, "and they'll o'ergrow the garden, / And choke the herbs for want of husbandry" (3.1.32–33). Cade, too, the text makes a special effort to inform us, eats grass and herbs (like weeds do) in the quiet walks of well-maintained gardens. The association of rebels and malcontents of one sort or another with flora and fauna is repeatedly reinforced throughout the play: they are variously described as drooping corn (1.2.1–2), "a limb lopp'd off" (2.3.42), a droopy pine (2.3.44), a raven in dove's clothes (3.1.75–76), a wolf in lamb's clothes (3.1.77–78), "blossoms blasted in the bud" (3.1.89), gnarling wolves (3.1.192), laboring spiders (3.1.339), starved snakes (3.1.343), ravens in wrens' clothes (3.2.40–44), "an angry hive of bees" (3.2.125), a kite (3.2.196), crab-tree fruit (3.2.214–15), infected air (3.2.287–88), lizards' stings, serpent's hiss, and boding screech-owls (3.2.325–27), "loud-howling wolves" (4.1.3)—all clearly not the favorites of plant and animal husbandry, not the features that are imagined to support the kind of order in nature necessary for the production of aesthetic and economic commodities. And by placing the human on the same level as the morally inconsiderable natural world, these metaphors implicitly carry possibilities and permissibilities for mortal violence in their meaning.

Moreover, in the abundant comparisons between rebellion and certain images of nature, we can see that "correspondences with the symbolism of popular culture are," as François Laroque maintains, "deeply embedded in the imagery of the play, which insists so much on the parallels with the animal world that it is often close to a fable" (82). The cultural tradition of popular protest Patterson talks of sees the natural world as a kind of mirror for the privileged image of human subjectivity. It is here, within the space of human subjectivity, that the important things happen. Nature merely reflects, confirms, or opposes those things.

What until act 4, scene 2 (the introduction of Cade) has at least some promise of potentially subversive drama[2] dilutes into comic carnivalesque inversions that are contained, doomed to reaffirm the order they oppose, by trivializing their own positions. Jonathan Dollimore seems to argue against any notion of absolute containment, claiming that "to contain a threat [to social order] by rehearsing it, one must first give it a voice, a part, a presence—in the theatre as in the culture. Through this process the very condition of something's containment may constitute the terms of its challenge" (xxi). The problem here is that not all voices are the same, and the *kind* of voice must surely determine the subversive potentials of the drama. Louis Montrose affirms a position similar to Dollimore's, arguing that "it is usually the case that the end of the play serves to reaffirm the dominant positions; nevertheless, the prior action may have opened up challenges and alternatives that subsequent attempts at closure cannot wholly efface" (123); but again, the question about tone is not addressed here. Exactly what kind of subversive moments are opened up by Cade? How badly does his spurious logic about francophones being enemies[3] compromise his voice as a rebel with legitimate grievances? How does his promise to make "the pissing-conduit run / nothing but claret wine" (4.6.3–4) erode his appeal as an activist and cast him into the wholly ineffective lunatic fringe? How do his self-confessing asides to the audience that he is a liar about his lineage blunt any edge of subversive authority he might have had? And from the heavy capriciousness of the mob that trails Cade, to Dick the Butcher's slapstick asides following Cade's every comment in act 4, scene 2, the "rebellion" is as far from seriously subversive as pigeons are from scholarship; but this does not mean that Cade and his crowd are a harmless bunch. "Sharp weapons in a madman's hands" (3.1.347) represent a serious situation—and the seriousness of the Cade rebellion is not ideologically dispelled by the simple dispersal of the crowd. *2 Henry VI* takes pains to take care of Cade, significantly in a manicured garden which, far from being an ideologically neutral space, is established as one that will not host chaos and disorder by unchecked nurturing of a rebel. The garden is nature stripped of its own order: it is power over nature materialized. The order that is imposed on what is repeatedly conceptualized in the early modern period as unruly, chaotic, and threatening nature is also imposed on Cade, the wild limb that is lopp'd off, the "trunk left for crows to feed upon" (4.10.84).

Ideologically, the production of the garden as a controlled space is part of a continuum of control: violent assertions of power over sedition and over imagined social disloyalty are also a part of that continuum. Far from being festive, the carnival atmosphere that blows through so much of this play finally rests on severed heads and puddles of blood.

VEGETARIAN ETHICS

Because the inversions that characterize carnival temporarily resituate "the natural" from its position in what Montrose calls a "coercive hierarchical model" (122), carnivalesque topsy-turvydom of class, gender, and race is important for theoretical discussions about how we conceptualize the natural world.

Obsessed with the senses (and with feeding), carnival is an ecocritical issue that relates to issues of race (Caliban is evidently vegetarian), gender (patriarchies co-locate women and meat), class (historically, meat is expensive), and so on; moreover, it would be interesting to see a study that traces how a play such as *2 Henry VI* both participates in and subverts a popular radical vegetarian environmentalist ethic. How, for instance, is the compelled vegetarianism of Cade characterized as the diet of losers? Is this vegetarian in the garden of Iden in any way a comment on or critique of the vegetarians in the Garden of Eden and their rebellion? Exactly what *are* the implications of the correspondence between the butchery of people and the butchery of animals in this play, with Dick the Butcher erasing the boundary between the two and the king himself pleading for animals?[4] Though the king makes such a plea, the overall action of the play works to contain this subversive thinking in the very character from whose mouth it came.

Henry is a weak king, and his weakness is ideologically inseparable from his expression of sympathy for animals. If we recognize meat "consumption...to be the final stage of male desire" (Adams 49),[5] the king's lack of virility and potency, neither of which come off as desirable, taint and are tainted by his animal rights sympathies. The subversive promise but ultimate containment of the play's critique against meat is part of a larger tradition that silences popular radical vegetarian environmentalist ethics, ethics that find spectacular expression in the work of Thomas Tryon.

Although hardly the "fervent exponent of vegetarianism" Andrew Wear claims him to be ("Making Sense" 129), Tryon certainly does take an ethical stand on the consumption of animal flesh, maintaining that "there is greater Evil and Misery attends Mankind, by killing, hurrying, and oppressing his Fellow-Creatures, and eating their Flesh, and that without distinction, than is generally apprehended or imagin'd" (232, $Q_4{}^V$).[6]

Tryon gives a long, detailed, sometimes redundant argument listing twenty-three reasons why people should not eat meat. Among Tryon's arguments are that "flesh and fish cannot be eaten without violence" (233, $Q_5{}^R$); that violence against animals proceeds "from the very same Root [as] does proceed all Back-biting, Envy, Strife, Rancor, and Contention" (234, $Q_5{}^V$); that meat-eating goes against the original plan of Judæo-Christian mythology (237, $Q_7{}^R$); that meat is not necessary for the human diet; that it is bestial for people, who have the intelligence to choose what food to eat, to choose to eat flesh like beasts; that it is less appetite than lusting for power, a "Dominion in Man, over the meek...[that] could not be satisfied" (249, $R_4{}^V$) that causes humanity to eat meat; that flesh is dirty and defiling; that flesh is less healthful than vegetables and causes "clotty blood," "hard Lumps," and "fills the body with abundance of slimy, corrupt juices" (or, in modern parlance, is high in cholesterol, causes cancer, and encourages acne—270, $S_7{}^V$); that "flesh and blood is too near of kin to the animal life in Man" (270, $S_7{}^V$) and that it is unseemly and unnatural for people to "eat the flesh of their fellow creatures" (271, $S_8{}^R$); that meat-eating proceeds from the spurious assumption that if we did not kill and eat animals, then we would be overrun with them;[7] and that it is disgusting.[8] Tryon is as concerned about the ethics of causing suffering as he is about health and spiritual matters, and he frequently talks about the pain that is involved for the animals in the production of meat and animal commodities. For instance, he talks about how the practice of refraining from milking cows before bringing them to market for sale "put[s] the creatures to much pain" (326,

$Y_3{}^V$), and throughout his discussion insists on the biblical axiom that people do to others as they would have done to themselves.

What is truly astonishing is that in the following chapter, Tryon goes on to offer a voice for "the Dumb: Or, the Complaints of the Creatures, expostulating with Man, touching the cruel usages they suffer from him" (333, $Y_7{}^R$). It is embarrassingly anthropocentric, but it is well-intentioned. Tryon voices the complaints of cows, oxen, sheep, and horses. At times, these complaints seem almost self-parodying: "Cruel and hard-hearted Man!...We COWS give him our pleasant milk; which is not only a most sovereign food of itself, but, being altered, and variously dressed, makes a great number of wholesome dishes; but this will not content them" (334, $Y_7{}^V$—336, $Y_8{}^V$). Arise, fair sun, and kill the envious moo! Well-meaning though it might be, speaking from the mouth of a cow seems ridiculously ineffective and undoes the authority of the very sensibly argued previous chapter.

Although it is beyond the scope of my inquiry here to chart meaningfully the place of vegetarian ethics in popular protest movements in the early modern period, I would be remiss not to mention that vegetarian ethics, despite their containment in 2 Henry VI and Tryon's A Way to Health less than a century later, was a live issue that both preceded and followed Shakespeare, especially with the rise of scientific medicine, and with meat increasingly being associated with questions of disease.

PART 2: ILLNESS AND ENVIRONMENT

Early modern disease in contemporary critical commentary constitutes an enormous and growing topic, yet with the mass of writing that has been done on it, there have been no analyses of relationships between illness and social protest, both of which were increasingly being discussed in environmental dimensions in the early modern period.

Although the topicality of disease and illness in the early modern period cannot be overstated, there are numerous perils attending attempts to historicize early modern literary representations of illness. The gap between literary representations of illness and the experience of disease by early modern people is a difficult one to bridge, and attempting to extrapolate intelligence about the material experiences of the ill from literary representations of disease seems to me misdirected, given the heavily metaphoric, analogic, and thematic positions disease occupies within literary texts. My purpose here, therefore, is not to glean information about real people suffering real diseases from Shakespeare's sick people; rather, my interest is in how, as a process that conceptually links human and natural, representations of illness express experiential reckonings with the stuff of nature, stuff that at times authorizes, licenses, and validates forms of resistance and subversion, as we will see.

Thematic discussions about environment and disease in literature are fairly uncomplicated; such discussions are, moreover, among the necessary methodological avenues down which ecocritical readings must, at some point, travel. It is gratifying work in that it gives the sense of accomplishment that often comes with quantitatively measurable results; however, catalogs of image patterns, discussions about disease metaphors, and simplistic attempts to correlate literary symbolism with "historical" events, while gratifying, will, unfortunately, not take literary theory and criticism beyond the boundaries of political insularity and detachment, boundaries which have often kept English literary studies from realizing radical potentials in the ways that other disciplines (such as geography) have begun to realize theirs. For ecocriticism to continue to cross such borders, however, is to take those crucial toddler steps, to make those readings which look for themes, images, metaphors, symbols, and so on, and from there to follow up, to look beyond how the natural environment figures in literary texts to how *it* is figured, how, within discourse, it is understood, what biases are written against it, how such biases both subvert and maintain structures of social and political power in given texts, both conceptually (in the metaphors through which texts understand the natural world) and materially (in the prescriptions about the ways in which to live through it).

One of the reasons that the literate understanding of illness in the early modern period deserves attention from an ecocritical perspective[9] is that in the representations of illness, we witness a linking of two worlds (the human and the natural), realms that in other discursive areas of the period are increasingly being drawn apart. Illness in the literature of the time is often a process of hybridization of the human and the natural, a process and experience of blurring, rather than a static identity of dis-ease that threatens the firm division between human and nonhuman.

BACKGROUND

Beliefs that there are relationships between the natural world and the health of its inhabitants have, by the early modern period, long been held. Mary Dobson explains in her meticulous examination of the topography of death and disease from the early modern to the modern periods that "for at least five millennia, men and women have observed and recognized that patterns of sickness vary according to locality and season, and that certain attributes of weather or the environment might be related to fluctuations and variations in ill-health and well-being" (1). She goes on to argue that this recognition registers both in the language people have used ("feeling under the weather") and in various therapies prescribed for illness (going to spas, getting some fresh air, etc.). By the time of Shakespeare, illness is a constant concern.

Fears about "the plague" and about plagues in general plague the early moderns. This is not really very surprising when we consider the enormity of the situation. One contemporary author reckons that the first wave of the plague, which occurred in the 1340s, had "a probably average mortality of around 47% or 48%" (Horrox 3). There are inklings about the causes and cures, but there is really no solid understanding of either. For the most part, it is baffling, and there are no clear environmental causes that can logically be discerned. As Dobson explains, the "elusive patterning of epidemic visitations along the divides of airs and waters, across the seasons of want and plenty, against the directions of winds and weather, puzzled and frustrated physicians in their search for simple environmental causal associations" (21). Nicholas Bownd's 1604 comments about how the plague "is so scattered" across the land (43, G_2^R) and his questions about why "this pestilence is so greatly in one part of the land and not in another" (67, K_2^R) are representative of the sheer confusion wrought by the plagues that blew through England every so often.

By the beginning of the seventeenth century, documents are regularly appearing in England that link environmental health and human illness. Thomas Dekker's *Newes from Graves-end* (1604) claims that "a plague's the purge to clense a cittie" (F_4^R) of the filth that accrues from overpopulation, from "having too too many, living / And wanting living" (F_4^V). Dekker had a clear understanding that when the carrying capacity had been passed, natural population checks would activate:

> When fruites of wombs passé fruites of earth
> Then famines onely physic: and
> The medicine for a ryotous land
> Is such a plague. (F_4^V)

There is also, despite the confusion wrought by epidemics, a basic recognition of the need for maintaining good sanitary conditions, of which Thomas Cogan is representative when he argues in 1584 that people must

> be circumspect in approching the persons who have hadde the plague, and much more in entering into the houses that have beene infected: and most of all in touching the clothes of those persons or places where the plague hath beene. For these thinges retaine the infection longer than the ayre it self. For in those persons that have beene infected, the poyson remayneth the space of two monethes. The houses and the householde stuffe, unlesse they bee purified with fire, perfumes, washinges, and such like, keepe their venom for the space of a yeare or more. (265, $L_1{}^R$)

The way Cogan conceptualizes contamination here is fairly sophisticated in that it distinguishes air-borne from material contamination, but the recognition of the importance of the former issue—air quality—in the state of people's health was common.[10] There was a common-sense understanding that, as James Hart explains in 1633, "being corrupted, it [air] proveth often the cause of many diseases" (14, $C_3{}^V$). Thomas Lodge makes a similar argument about uncleanliness and corruption, maintaining that plague is caused by

> the indisposition of the earth overflowed with too much moysture, and filled with grosse and evill vapours, which by vertue of the Sunne beeing lifted uppe into the ayre, and mixed with the same, corrupteth the nature and complexion thereof, and engendreth a certaine indisposition in the same contrary to our substaunce, from whence it commeth to passe, that they who sucke this infected aire are in daunger to be attainted with this contagion and sicknesse of the Pestilence. ($B_4{}^{R\text{-}V}$)

For someone such as Lodge (writing in 1603), it is clear that "the plague is a contagious sicknesse," which can lurk in unclean sheets, pillows, blankets, and so on ($L_2{}^R$), but he does not know how or why. For someone such as Bownd (a year later), the why is clear enough: it is God's wrath that causes the plague;[11] but for the more secular Lodge, the causes are clearly more environmental, while it is the *cures* that are within the purview of the divine. People "may by Gods helpe, and by keeping good order, auoyde the plague" ($L_3{}^R$). Since disease, illness, and plague are often, as I discuss further below, understood *as* chaos, disorder, and loss of control at the time,

Bownd's writing is clearly tautological: "avoid disorder to avoid disorder" does not clarify things, and, anyway, as Science was to find out, there were far better cures than religion. Nevertheless, though the kinds of knowledge that were to become available with the development of germ theory are clearly not available at this time, there *was* often a basic understanding of the general concept of infection, an understanding that was frequently expressed in terms of humoral theory. Cogan, for instance, believes that "the Sanguine sort are soonest taken with this infection, and next to them the Cholericke, thirdly the Flewmatike: and last of all, the Melancholike: because the colde and dry humour is least apt to inflammation, and putrefaction" (273, Mm^R).[12]

By 1676, John Graunt is writing extensive and detailed studies relating deaths to environmental influences in *Natural and Political Observations,* and by the end of the century (1697), Thomas Tryon is explaining "Why Cities and great Towns are subject to the Pestilence and other Diseases more than Country-Villages" (169, M$_5$^{R}).

With his characteristic logic, Tryon talks about overcrowding and the human waste pollution that results; pollution from "smoke, dust, and ashes of sea-coals" (170, M$_5$^{V}); pollution produced in the slaughter of animals; and so on. Many of these are, in Tryon's thinking, natural poisons—"Poysons in Nature" (170, M$_5$^{V})—that are occasioned by humanity and increased beyond their natural levels to the point "that they overcome the pure Vertues" (170, M$_5$^{V}) and cause illness. Tryon's interest is in tracing both material relationships between society and nature and the far less material relationships described in correspondence theories which insist that nature provides "a mirror image of human social and political organization" (Thomas 61). The correspondence is sometimes very direct for Tryon, as is clear when he imagines that

> if the people of any Town or City give way to uncleannesses in Meats, Drinks &ct. and addicted to the Impurities of Venus, as most such places are, then by mutual Inclination the like Property in the Coelestial Bodies and Elements are excited, and by degrees contaminate the whole atmosphere (or the parts of the Air next the Earth) with Pestilential Poysons, causing Botches, Boils, Venerial Diseases, Fevers and Plagues, all according to the degree of the awaken'd Wrath, and the length or shortness of the time of its operation. (170, M$_6$^{V})

Even with this kind of correspondence theory, though, the interest with Tryon is in causal material relationships. Before Tryon and Graunt, there is very little understanding about the causal material relationships between social issues and environmental forces that together produce or encourage illness in early modern England, though there is substantial speculation. Infection is the sine qua non of illness in much of the literature here, and the susceptible bodies are understood to be those in which there is an imbalance of humors. As Gail Kern Paster explains, "people imagined that health consisted of a state of internal solubility to be perilously maintained" (8). It is a view that was on its way out.

Atomization and mechanization in the early modern period meant that in the emerging medical discourses, there was more of a tendency to offer "causal accounts of illness based…on localized events taking place within the body rather than on a generalized imbalance of the humors" (Wear "Explorations" 118). The movement away from humoral theory is also, in a sense, a movement toward understanding external events acting on the body and causing internal dis-ease. Still, the transition was far from complete, and there were "problems associated with fitting new explanations of disease into the framework of Galenic medicine" (ibid.). It is within this context that we find the metaphor of illness in *2 Henry IV.*

2 HENRY IV, FALSTAFF, AND ILLNESS

Thematically, disease is central to all of the action of *2 Henry IV*, and images of illness fill the air from beginning to end. Sickness is implicit at the very start of the play with the personification of Rumor. The images Rumor conjures as it boasts about spreading through the land like a plague, implicitly connecting disorder with a fearsome natural world and dropping into people's ears from east to west, might well be those that Disease personified would use. The association of disorder with the nonhuman is quickly reinforced in the dialogue that follows between Lord Bardolph and Northumberland concerning the chaos that is spreading like unloosed horses.

The king is sick, and through his jaundiced eye, the natural world becomes an inhospitable place. The sea is a place of "rude imperious surge" (3.1.20) and "ruffian billows" (1.22). The hours are "rude" (1.27), the clouds "slippery" (1.24), and sleep, "Nature's soft nurse" (1.6), elusive to this sick insomniac. It is as though the king in his sickness sees a reality that all young and healthy people, blinded by their good fortune, do not see. He claims that

> if this [what he imagines he sees] were seen,
> The happiest youth, viewing his progress through,

What perils past, what crosses to ensue,
Would shut the book, and sit him down to die. (3.1.53–56)

Though the king is the central site of disease, from Rumor spreading through the land to Falstaff urinating in a bottle for a doctor
(1.2.1–5), disease is a reality that must be dealt with. As a metaphor,
however, it is qualitatively different from the main metaphor that runs
through a play such as *Hamlet*, where rot and excess are key issues. It
is true that in both *2 Henry VI* and *Hamlet*, the social system is not
functioning properly, and that excess is the cause of the problems in
both plays, but in *Hamlet* we are well beyond disease and into putrefaction. It is precisely this kind of decay, such "rotten times" (4.4.60),
that the old King Henry fears and envisions as the unwholesome
progeny that will inherit the future: he tells the Dukes of Clarence
and Gloucester,

The blood weeps from my heart when I do shape,
In forms imaginary, th' unguided days
And rotten times that you shall look upon,
When I am sleeping with my ancestors. (4.4.58–61)

He fears Prince Henry is following a lavish and overindulgent lifestyle,
one that can only lead to satedness, rot, and stupor, and that Henry
will be unable to leave his life of excess and consumption: " 'Tis sel-
dom," the king claims, "when the bee doth leave her / In the dead
carrion" (4.4.79–80). As it happens, Henry does leave his naughty
ways, and all is healed by the end—except for the old king, who is
dead. On a thematic level, there is not really very much more to say
about disease—at least, not much that will contribute to an ecocritical understanding of the ideological ties that bind illness, resistance,
and environment with each other in this play.

But as a metaphor for social unease central to *2 Henry IV*, illness
becomes a materialization of chaos, and though it is not quite understood, it is feared and fought against because it threatens the very
existence of the State. The king is sick, and the body of the "kingdom, sick with civil blows" (4.4.133), foul with rank diseases that
grow near the heart (3.1.38–40). Rebellion here, as with the real or
imagined social disloyalty in *2 Henry VI*, is always associated with the
natural environment. And loveable though he is, Falstaff—coarse,
bestial crossover to the natural world—represents and embodies a
serious threat of rebellion, less by his animalism than by his partial
admittance to the jurisdictions of civic power. He is a threat against

both the order that the young king will embrace and against the progress away from Galenicism.

Falstaff reads Galen (1.2.117), and his sherris speech (4.3.115–25)— where he talks about blood and fluid temperatures as the cause of Henry's problems—is terminologically Galenic. Falstaff is the residual, both within the play and within the changing medical discourses beyond the play.[13]

It is always difficult to assess the relationship between a text and the context out of which it grows and to which it speaks, but in this play, Falstaff offers a good link. Within the fictive reality of the storyline, the old king is real sickness, but Falstaff is clearly metaphorical sickness, the greatest parasitic infection that threatens the young king. It is startling to hear loveable old Falstaff say to his slow-witted friends "Let us / take any man's horses, the laws of England are at / my commandment. Blessed are they that have been / my friends, and woe to my Lord Chief Justice" (5.3.136–38).[14] Since he has nowhere shown himself capable of being anything but an irredeemable drunkard—albeit, a humorous and endearing one— concerned only with satiating his own maw, there is no reason not to take him at his word. He hopes, in his protocapitalistic way, to cash in on the diseased state of the kingdom, to "turn diseases into commodity" (1.2.248). He is a potently subversive figure, not in any politically progressive sense of revolting against authority for the common good, but in the manner of individualism run wild. In one sense, he is a man of his time, knowledgeable about the fact that disease is a commodity through which practitioners of health and well-being ply their trade and self-centered in all the ways that allowed such trade to develop into full-fledged capitalism. In another sense, however, he is a relic, a dinosaur, a thing of the past that cannot keep up with the very rapid changes that are occurring around him, and this is his tragic flaw, one that is, to be sure, maintained by his acute individualism. He is at one and the same time posited as a thing too, too natural, too far removed from the civil and the human and at the same time is a thing too, too unnatural, a freak that does not evolve (and the concept applies even though we still have some time to go before we reach Darwin). He is a disease, a blot on the face of the natural.[15] He ignores change. This, according to the text, is unnatural, and it is a disease: "It is," Falstaff asserts, "the malady of not marking, that I am troubled withal" (1.2.121–22). Time and change, however, go on regardless of Falstaff. And what we are talking about here is *not* a development or evolution from the past but a paradigm shift from it, a betrayal of it, the new of the old, which the old king fears is going too far.

He seems to fear a paradigm shift run amuck, with the son betraying the father, and all apparent continuity being tossed to the winds, scattered in disarray and disordered like "the times," which Northumberland tells us at the beginning of the play, are out of control, are a chaos spreading (diseases also spread) like wild horses (1.1.9–11), casting the human into the disorderly realm of the natural and the wild, a realm of the rebels, who, under the archbishop of York and upon hearing of the peace-deal that has been struck, "Like youthful steers unyok'd,...take their courses / East, west, north, south" (4.2.103–4). The king fears that the betrayal that hangs so heavily over so much of the action of the play will finally drop down on and kill him, a filial ingratitude he can only conceive of as monstrous, something that goes against nature: "How quickly nature falls into revolt," the king complains, "When gold becomes her object" (4.5.65–66). It is a kind of betrayal that, in his view, "conjoins with my disease" (l.64).

If anything, however, his son's practice of betrayal is moderated to the level of the remedial and is not complicit with disease. Like any other remedy, too much of this one will kill the patient. We have been prepared since act 1, scene 1 for precisely this close relationship between poison and physic, when Northumberland, bemoaning the news of the death of his son (Hotspur), says, "In poison there is physic, and these news, / Having been well, that would have made me sick, / Being sick, have (in some measure) made me well" (1.1.137–39). He goes on, a dramatic second or two later, to mention how disease *is* chaos, fires out of control, wild floods drowning order, to use his metaphors, which are the work of, to use his words again, "Nature's hand" (l.153).

In calling upon nature, the king and Northumberland both remind us of how heavily entwined illness and the natural environment are in this play, and how inseparable are the representations of illness from those of political instability.

The play's representations of human problems (such as illness and social disorder) as effects of environmentally produced malevolence, representations at once anthropocentric and bestializing, mark a fear and abhorrence of the natural world that inspires the dying king to remark that his kingdom "will be a wilderness...peopled with wolves" (4.5.136–37). It is fear and abhorrence of the natural world that we hear written into Gloucester's talk about "loathly births of nature" (4.4.122), which itself plays into a larger largely ecophobic deliberation occurring in the early modern period on the question of monstrosity (see chapter 5 below). And, ironically, the space that we might expect to be least contaminated by the ills that plague the court, the space of nature, like *2 Henry VI*'s garden of Iden, far from being idyllic and unturmoiled, is a space of social competition. It is in the forest of Gaultree that *2 Henry IV* gives us perhaps the play's most shocking display of bad faith and betrayal, with Prince John arresting Scroop, Mowbray, and the rest of Hal's opponents who are present.

To find an unturmoiled space in *2 Henry IV* or *2 Henry VI* is impossible, and nature is coded heavily in each. Calibrated by the central preoccupations of social unrest and sickness that beleaguer the Elizabethan imagination, the environment offers a vast resource through which the plays define social and physical dis-ease, and through which these in turn define the environment. The ideological work of boundary erasures that *2 Henry VI* and *2 Henry IV* perform—and ironically at a time when such boundaries are elsewhere being so vigorously asserted—raises important questions both about the conceptual relationship between protest and environment and about the social production of "natural" space. The illness and social decay of *2 Henry IV* and the ambition and rebellion of *2 Henry VI* are images of disorder that place rebels outside the realm of the moral consideration that at the time was accorded only to unambiguously human subjects. Dangers to the State no less threatening than weeds are to grain and sustenance and disease is to health and well-being, rebels, denied the status of human subjectivity in these plays, succeed at least in challenging boundaries, regardless of how poorly they may fare at being rebels, and regardless of how well the plays fare as subversive drama.

But they do something else: they suggest possibilities for ecocriticism far outside of its traditional grounds. For an activist scholarship, class becomes the main concern of a play such as *2 Henry VI*, in which we find rebellion front and center. Increasingly, it seems untenable to discuss class meaningfully outside of an ecocritical framework, since so very much of our thinking about class and social hierarchy is structured by ecophobia, by the way we lay value on, commodify, and hierarchize nature.

If ecocriticism is useful for analyzing class struggle in *2 Henry VI*, it is no less so for a play such as *2 Henry IV*, where disease seems the central thematic preoccupation: for this play, ecocriticism is useful because it helps us to see and explore ecophobic links that connect chaos and nature, disease and social unrest, environmental ethics and community. For an activist ecocritical scholarship, seeing these links in unfamiliar but, paradoxically, profoundly influential literature is an important beginning.

5

Monstrosity in *Othello* and *Pericles*: Race, Gender, and Ecophobia

> It is not enough to claim that human subjects are constructed, for the construction of the human is a differential operation that produces the more and the less "human," the inhuman, the humanly unthinkable. These excluded sites come to bound the "human" as its constitutive outside, and to haunt those boundaries as the persistent possibility of their disruption and rearticulation. (Judith Butler, *Bodies That Matter* 8)

In part a result of the ecophobic genuflex evoked by allusions to monsters and monstrosity, the haunting of the boundaries about which Judith Butler speaks recurs along several different—sometimes intersecting, sometimes confluent—axes in plays such as *Othello* and *Pericles*. The plays' obsession with monsters jiggles orders, hierarchies, values, rules, and forms defining nature. These definitional axes of nature (the product of long cultural struggles), while neither firm nor finalized, and while always contested, are also always—in some sense—controlled and always predictable. It is when they are not controlled and predictable that problems arise. Writing monsters imagines unpredictability and agency in nature and opens a space for a variety of discursive disciplinary actions against such imagined unpredictability and agency. Written as the source and guarantee of behaviors and relationships (sexual, psychological, and racial) among people, nature—imagined as devoid of sufficient agency to discipline itself, to deal with the monstrous transgressions it has itself spawned—must be (and invariably is) disciplined. This chapter discusses in detail the types of disciplinary reaction—the types of ecophobia, in other words—that the monstrous imagination invokes along the axes of race and gender in *Othello* and diet in *Pericles*.

CONCEIVING AND DISCIPLINING
THE BODY: A CASE STUDY OF *OTHELLO*

Georgia Brown has written eloquently about the doubt and skepticism monsters engender: "Monsters deal in doubt and the difficulties of articulating doubt [...]—doubt about anxieties, about what is natural or unnatural in humans, about what we know and do not know, about the efficacy of language" (57). Although the argument is convincing as far as it goes, it is perhaps more than simply doubt that monsters evoke: they present the horrifying aspect of an agential nature that helps codify and organize rituals of scapegoating on the one hand and the parameters of exploitation on the other, while at the same time feeding a felt hunger for wonder.

Central to the articulation of boundaries monsters cross and help create is "the anatomy," literal and figurative. The importance of anatomies in the early modern period has been well documented and discussed. Devon Hodges has argued that the anatomy lesson is a kind of experiment in deconstruction, a "method that turns bodies into parts" (17), undertaken in a "passionate effort to get back at a solid unified truth" (15). It has everything to do with power—the power to define the limits of the human, the limits of border-crossing, and difference. It is a power that is exercised through the knife, through excisions and divisions: "To know a body...is often to dominate, conquer, master, discipline, and punish it. The science of investigation and surveillance...makes the body an object of knowledge by placing it within a controlled order and separating it into individual elements" (5n12). The psychology of mapping here has implications that move far beyond the body.

When Mary Douglas claims that "cosmologies cannot rightly be pinned out for display like exotic lepidoptera, without distortion to the nature of a culture" (*Purity* 91), we understand that indeed any map—whether of a subway system, a culture, or a body—is a rearticulation, distortion, and disciplining of the component parts and their relationships. Moreover, cartographic and anatomical metaphors that enable and encourage ways of seeing parts separated from wholes carry violence. Dismembering specific social practices from the cultures of which they are a part cuts such cultures in much the same way that discourses about body parts implies violence to the integrity of the body. The effect is a powerfully incisive and dishonest cultural fashioning.

Often, the violence that the anatomy lesson performs on the body is linked with the violence written into early modern notions of cultural and environmental difference. Bodies such as Othello's are written less as unified wholes than as collections of dismembered parts.[1] The body of the racial or cultural figure of difference—so often confronted in the early modern era—is both discursively and visually disarticulated, anatomized as objects that share the same ontological status as the natural environment. Hence, the heads lying around in Plate XXII of Theodore de Bry's *Americae, Pars Quatra*, come not from subjects but from objects, parts of the natural environment ripped from the secure ontology of the human in a preemptive disciplinary slicing, a surgical removal of the threat.

Othello, ontologically associated with the "rough quarries, rocks, and hills whose heads touch heaven" (1.3.141) in the tales he uses to woo Desdemona is, by any account, a monster. Karen Newman describes him as "a monster in the Renaissance sense of the word, a deformed creature like the hermaphrodites and other strange spectacles so fascinating to the early modern period" ("Wash the Ethiop" 153). Significantly, the discipline that the text exerts on the dangerous Othello is at one with the excisionary discipline of the anatomy lesson—he is cut off of the body politic, like a wart from a foot. Moments before, in his "here is my butt" (5.2.267–282) speech,[2] he melodramatically begs for but does not receive corporal discipline: "Whip me, ye devils, /.../ Blow me about in winds! roast me in sulphur! / Wash me in steep-down gulfs of liquid fire" (5.2.277–80), he effuses with great histrionics. He makes one last self-dehumanizing gesture—"I took by the throat the circumcised dog / And smote him—thus" (5.2.355–56)—and kills himself.

Situating the dramatic dehumanizations *Othello* constructs within the framework of the discourses of monstrosity that spectacularized corporeal difference helps us to understand the *struggles* the early moderns had in defining the precise boundaries of nature. Ambroise Paré's *On Monsters and Marvels* is a document of such struggle, one that registers as much fear as certainty. While the " 'thick descriptions' of Elizabethan culture and society that come to mark the type of inter-textual analyses generated by New Historicists" (Hendricks 4) shed an enormous amount of light on how the psychology of superiority is overwritten by radical insecurities, by fears of uncertainty, by doubts, and by phobic responses to imagined unpredictability, drawing clear causal relations between something such as Paré on the one hand and *Othello* on the other seems a delicate business at best. The fact, though, that *Othello* participates in a discourse that is beset with insecurities and neuroses is itself a comment on the implications and complexities of the ethics here.

Othello's suicide positions the eponymous hero within a discourse at once racial and monstrous. The comparison Othello draws between himself and the "turbaned Turk" (l.353) whom he took by the throat is invested with a heavy interest in the body. The throttling here is important, as are both the animal imagery and the phallic mutilation in Othello's language. The images of dismemberment here are heavy. Although a psychoanalytic reading could argue that Othello's language suggests some anxiety about the head of the penis, some castration or circumcision anxiety,[3] and while in the comparison to the throttled dog we can perhaps hear Othello saying that the logic of his predicament strangles him, it is clear enough that the audience imagining Othello must at least *think* of throats and hypoxeiated heads when Othello kills himself. It is also clear that when Othello smothers Desdemona, he kills her by cutting off the flow of oxygen to her head. In both cases (and perhaps also in the case of the Turk's penis), there is a symbolic decapitation, either enacted or envisioned, not noteworthy in itself, but revealing in light of the way that Othello wooed Desdemona: he told her stories about men without heads, or, more accurately, about "men whose heads do grow beneath their shoulders" (1.3.144–45—see figure 5.1 below).[4] There is an implicit symbolic disciplinary dismemberment that results from Othello and Desdemona's love.[5]

His final lines, Patricia Parker has argued, "suggest a proliferating series of exoticized others" ("Fantasies" 98) and ends in his suicide. The text kills him, but this excision is ambivalent.

Othello is himself an ambivalent figure, at one and the same time in the heart of *and* outside of—under the foot of—the Venetian power structure. The tragic dimensions of the play are possible only to the extent that Othello is simultaneously humanized and bestialized. It would be a mistake, though, to argue, as Eldred Jones does, that what we see in Othello is the "complete humanization of a type of character" (109). Rather, the opposite is the case, since the conceptual linking of women and people of color with animals is an implicit debasement from human to animal status, a *de*humanization, in other words. If we accept the thesis that "the body is properly human only when it is culturally acceptable" (Currie and Raoul 18),[6] then Othello is not properly human. Othello is not culturally acceptable. For one thing, he is black, and for another, Desdemona is not.[7]

There is nothing ecologically innocent about racism: plants, animals, time, and people are alike cut with the same knife, producing a bidirectional assault—one leveling people to the same sphere of

(A)

Figure 5.1 Men whose heads do grow beneath their shoulders.

(B)

(C)

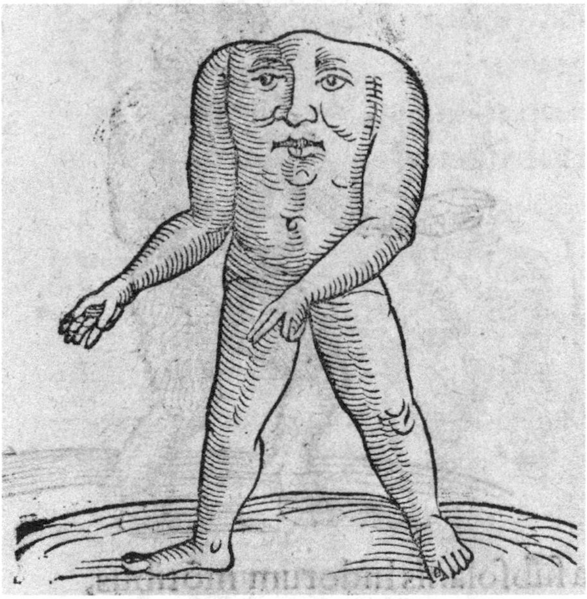

Figure 5.1 Men whose heads do grow beneath their shoulders (Continued).

Source: Figure A is from Sebastian Münster *Cosmographia* (1554). Figure B is from Hartmann Schedel's *Liber chronicarum* (1493). Figure C is from Konrad Lykosthenes's *The doome warning all men to the iudgemente* (1557).

moral considerability as the environment, the other ascribing a set of values to an insentient and neutral nature. Much of the villainy and otherness depend for their full effect in *Othello* both on the bestial-izing of Othello (and his implicit and explicit associations with the natural environment) and on the presence of an essentialized under-standing of color, a chromotypic essentialism where light is good and dark is bad.

The meanings immanent in the rhetoric of "black" and "white" and the racist epithets hurled around in *Othello* (and certainly in other plays of the time, *The White Devil*—discussed further in chapter 8 below—being a good example) contribute to the presentation of pre-packaged, essentialized identities, in which lasciviousness, jealousy, and bestiality come as part of the black package. Othello is black and therefore credulous,[8] easily made jealous, and before long he "breaks out to savage madness" (4.1.55)—because he is black. This discursive alignment of Othello with madness powerfully extends his distance from humanity proper toward a conceptual space of chaos, beyond the patternings of human power and control, a positioning consonant with dehumanization and bestialization into spaces of unpredictable and terrifying Nature.

Shakespeare certainly suggests possibilities for a fully integrated Othello, but, perhaps, only because he is a good soldier, and, as such, he is a good asset to the state. His military prowess can be used. When benefits outweigh deficits, many things can be overlooked: "Race is," as Ania Loomba remarks, "negotiable" ("Color" 29). But even if Othello were properly human, even if his color were not a problem, even if blacks were *not* "associated in the popular imagina-tion with monsters" (Aubrey 221–22), and even if Desdemona's color did not matter—which is to say, even if Emily Bartels is correct in arguing that race is irrelevant in *Othello*—the play still presents the body of the Moor. A focus on the body is always a foregrounding of nature.

In Iago's racist opinion, a body is all that Othello *can* be to Desdemona. Iago reasons that "when she is sated / with his body, she will find the [error] of her choice" (1.3.350–51). The play seems partially to confirm Iago's position. Obviously Desdemona *has* made a fatal error within the ideological framework of the play by choosing "against all rules of nature" (1.3.101). Nature is the authority here, and written into this nature are Venice's racist values. What Greta Gaard calls the sexualizing of nature and the naturalizing of sexual-ity produces nature as a thing that does not sanction Desdemona's choice and Othello as an animal who does. Nature, a thing feared for its vile and dehumanizing effects elsewhere in the play, is here endorsed as an authority over what people do with their bodies—an authority that people ignore at their peril.

Associated with nature through the play's vigorous emphasis on his body and through his sexuality, both of which stand, to borrow a phrase from Gaard, definitionally opposed to reason, Othello wins Desdemona's heart precisely because she does *not* see him primarily as a body. She loves him *in spite* of his body. As Othello explains, "She had eyes, and chose me" (3.3.189), and, no doubt, she chose him, as he explains earlier, because "she lov'd [him] for the dangers [he] had pass'd" (1.3.187) in environments with which his identity is bound. Her failure to identify Othello in the way Iago does stands in stark contrast both to Iago's perceptions and to the ideology of the text itself.

The synecdochal identifications of Othello as "thick-lips" (1.1.66); the focus in general on the materiality of his black body, his "sooty bosom" (1.2.70), and his "gross clasps" (1.1.126); the discursive working out of a fantasy of a magnificently strong and exotic and therefore potentially threatening body that seems properly trained and tamed to fight the right wars for the right side, yet remains a monstrosity of Venetian society in its absolute disregard for the boundaries of the color bar in place for early modern bodies; the dramatic realization of the threat of the black beast, culminating first in a slap and then in the murder of a white woman[9]—each part of the process here is powerfully anatomizing, disciplinary, and constitutive of the terms of Othello's material engagement with the play's fictional world. In the world of the play, Othello is a threatening body. At one point, Brabantio orders that "if he do resist, / Subdue him at his peril" (1.2.80–81), and certainly the implicit image of nasal mutilation in Iago's comparison of Othello to asses "tenderly…led by the nose" (1.3.401) suggests a disciplinary imperative that works through the body (in this case, the nose). But it is the very constitution of Othello *as* body that functions most perniciously as a disciplinary maneuver against the black Other in this play. Othello is the black Other who is embodied, disempowered, and set outside the community of *men*.

Therefore, what makes a man in a play such as *Othello*?[10] How does manhood constitute or stand in for the category of the human? In what ways do issues of male sexuality in *Othello* distinguish the human from the monstrous? Embedded in these questions about the parameters of "the human" are dangerous questions about the parameters and reach of nature, of what falls within its purview, of what is to be expected from it, and of what is considered ethically appropriate when it is imagined to be out of order.

Othello's duplicitous nature, his status as "a dually constituted subject" (Boose 38)—his imagined involvement with the bestial and his participation as a fully enfranchised subject—is less a contestation, though, of the period's boundaries than a conning of them. Ultimately, the "fraud" is revealed, and *Othello* reinforces rather than challenges the boundaries.

In the meantime, Othello—evidently unable to play his part with consistency or reliability in the fiercely racist Venice—becomes, like Iago, not what he is: "My lord is not my lord" (3.4.124), Desdemona apologizes when Cassio stands "within the blank of his displeasure" (1.128). In a sense, one might argue along with James Aubrey that Othello carries the morally monstrous offspring of Iago. Iago does, after all, promise to "bring this monstrous birth to the world's light" (1.3.404); he does impregnate Othello through the ear (as Aubrey 236—and, before him, Coppélia Kahn 144—so aptly explains); and Iago *does* ultimately engender monstrosity in Othello. Feminized, Othello's body is a *terra incognita* on which competing definitions of humanity are mapped out. Although his sexuality is never overtly questioned, questions about acts of sex do come up. The body is always at the fore in this play.

Iago's crooked talk about straight sex, with his animal imagery and all the implied racism, reminds us in a way that few other texts do about the *sex* that the two lovers have. This goes a long way to show that "racism *is* obsessed with...the body" (Read 1258—emphasis added), especially when there are two bodies of different colors making sexual contact. But what is even more interesting is how male-male "sex" leads to the monstrous in the play.

In addition to the implicit and symbolic male-female (yet obviously same-sex) relationship between Iago and Othello, there is also some kind of sexual transgression in the alleged dream and behavior of Cassio. Kahn argues that "like the microdrama of the whole play, [Iago's telling of Cassio's alleged erotic dream] confounds heterosexual intercourse, as Cassio dreams he is making love to Desdemona, with homosexual intercourse, as he suggestively embraces Iago at the same time" (145). Othello's reaction? "O monstrous! monstrous!" (3.3.427). The imagined sodomy is a monstrous thought. Sexual contact between men was so anathema to the legislators of bodily conduct that it is hard to imagine Othello not reacting with some vehemence. Same-sex sexual relations are monstrous, go against nature, and are practically unspeakable in Othello's mind.

APPETITES

There is a wealth of documents about the early modern sexual imagination. The sexualizing of America, captured graphically in Jan van der Straet's "America" and discursively in innumerable accounts,[11] for instance, promiscuates the New World and its inhabitants—but it is not so much "free love" as comparisons to and accusations of bestiality that are often represented. Such sentiments underlie the claims of Pietro Martire d'Anghiera (Peter Martyr) about New World indigenes. Writing in 1515 about, among other things, the sexual practices of people in the New World, Martyr claims that "there is no sense of justice among them...They are *bestial* and vaunt their abominable vices" (Peter Martyr, cited in Honour 58—emphasis added). Vices notwithstanding, these cannibals (literal and sexual), *sell*. The Old World public is hungry for New World adventure, and it dreams about a dangerous land waiting to be tamed.

Othello is certainly aware of the appeal of the cannibal and of its spatial dislocation. In his successful bid to woo Desdemona, he talks about "the Cannibals that each other eat / The Anthropophagi" (1.3.143–45). Dislocated from the geography of the center to a geography of difference, they share no substantial dissimilarity from the

Figure 5.2 "America," by Jan van der Straet

"rough quarries, rocks, and hills whose heads touch heaven" (1.3.141) that form another part of the tale of marvels Othello uses to win Desdemona's heart and imagination.[12] Cannibals may be possessed by whoever lays claim to the land, as Caliban, the anagrammatic cannibal of *The Tempest*, a "thing of darkness" (5.1.275), is possessed by Prospero, but they remain geographically removed. Similarly, Othello's cannibals and Caliban are far removed from the imagined domestic spaces of each play and are less threatening than the unwitting cannibal of *Titus Andronicus*, who (in inadvertently eating her own children in the "at home" of the fictional world of the play) presents a picture that is truly horrifying and beastly.

The semiotics of cannibalism, one of the vitally overlapping areas between postcolonial theory and ecocriticism, has changed very little over the past 400 years. A 1995 article in *Time* reports that "human fetus soup" (Dam, Emery, and Lai 12) has become something of a delicacy in Shenzhen.[13] The report plays into what seems a renewed anti-Asian trend in the West (one has only to think of the anti-import messages in car advertisements in the West)[14] and situates the alleged dietary trend "out there" in an exotic geography. Similarly, James Pringle's report in the *London Times* (April 13, 1998) situates cannibalism in the isolated, sequestered, secretive Stalinist North Korea. Perhaps it is merely a coincidence that at the time of the Pringle article, there were increasing tensions between North Korea and the West. And perhaps it is just bad luck that despite the amazing information technologies we have, no one was able to snap a single shot of people eating people (nor were there any pictures among the rash of reports about cannibalism in North Korea that followed).

Cannibalism is something disgusting that is reported from "out there," not "at home." The urge to situate cannibalism into geographies of difference may explain why relatively little discursive attention is given to cannibalism that *is* domestic. There are, of course, exceptions, and some work has recently been done with the question of sacramental communion.[15] Most of the work with cannibalism, however, takes postcolonialist approaches that largely overlook interrelationships between ecophobia and colonialism.[16] Both ecocriticism and postcolonial theory stand to profit from looking at how the semiotics of cannibalism participates in the writing of natural environments: cannibalism is a race and environment issue.

Stephen Slemon's "Bones of Contention" comes close to discussing how what he calls "the discourse of cannibalism" (165) is significant to the writing of a hostile environment. Slemon argues that the discourse of cannibalism "necessarily designates an absolute negation of 'civilized' self-fashioning in a place that is no place, and is always 'out there'" (ibid.). It is an ecophobic fashioning in that it offers a demonized geography that is to be both feared and despised. In such a schema, as Slemon notes, both the land and the people threaten to consume the travelers (163). While Slemon is clearly aware of the spatial importance of the topology and of the fact that colonialist discourse articulates a "managed *difference* in the field of 'nature'" through the discourse of cannibalism (165—emphasis in original), the *significance* of environment as it is configured in the conceptualization of otherness here remains unattended.

Semiotically, cannibalism makes people beasts, associates them with a loathsome and terrifying nature that the early modern imagination preferred to keep separate from the human sphere. Ecophobia is written into the discourse of cannibalism. People who eat people are like animals very far removed (ethically and geographically) from the space of humanity: as Albany in *King Lear* puts it, "Humanity [that] must perforce prey upon itself, / [Is] like monsters of the deep" (4.2.49–50). Anthony J. Lewis is surely correct in arguing that "the identification of people with food [in *Pericles* is] the reduction of human beings to comestibles" (155), but it is also a "reduction" of human beings to the natural world, a "reduction" that overlooks differences between people on the one hand and floral or faunal commodities on the other, and if we see this implied in *Othello*, we see it explicit in *Pericles*, where we hear of young women—at the age of fourteen!—being "ripe for marriage" (4.Gower.17); where we hear of the daughter of King Antiochus, unnamed in *Pericles*, being described as a fractal commodity, a precious "fruit of yon celestial tree" (1.i.21), "a golden fruit, but dangerous to be touched" (1.28); and where we learn that this fourteen-year-old woman is "an eater of her mother's flesh" (1.1.130) and has been happily having an incestuous relationship with her father for some time before the action begins. It is, moreover, a gendered dislocation that we witness here: it is women here, not men, who become edible commodities.

PERICLES AMONG THE CANNIBALS

It is certainly a very weird play. Severed heads, more storms and shipwrecks than most readers can confidently count, the miraculous preservation of persons alive under water or dead and unburied on land, a denouement which mixes, if not hornpipes and funerals, at least brothels and betrothals, and a remarkably accident-prone protagonist. (Nevo 150)

Nature probably is not the first thing that comes to our minds when Pericles, Prince of Tyre, goes to the court of King Antiochus seeking the King's beautiful daughter, nor when he is confronted with the odd father/daughter relationship.[17] We are more likely to think about relationships involving child abuse than to muse on relationships connecting sexual and environmental ethics, but the discourses of cannibalism and incest blur the boundaries of "nature" and "culture" in this play. As Alexander Leggatt observes, "When incest appears in Jacobean drama, it is generally treated as a fundamental violation of nature" (167). And violations of nature have dire consequences in the early modern imagination.

Visual artifacts representing the imagined material consequences of transgressing the imagined boundaries of natural propriety can be found in the many pictorial representations of monstrosity in the early modern period (see, e.g., figure 5.3). Incest, of course, figures significantly in the literature on monstrosity, and it (like the cannibalism with which it is associated) raises questions about inheritance. Whether it is children eating their mother's flesh, or mothers who "eat up those little darlings whom they lov'd" (1.4.44), we have, as Constance Jordan insightfully comments, "a present generation consuming its future" (345), not to mention polluting the body politic in *Pericles*.

Antioch is a troubled place, an "earth throng'd / By man's oppression" (1.1.101–2), a place of pollution. This pollution scripted as incest/cannibalism, triumphs and presents, at least for a nonincestuous audience, a loathsome, horrifying, and disgusting place. The underlying concept flies in the face of what both early moderns and we today imagine to be the way of a natural order. But if the characters violate nature, nature also violates the characters in this play.

Spatialized and mapped to provide at times a residence for monstrosity and at others an escape route from it, competing geographies flash through the play like a surreal slide show. Many critics have observed this barrage of scenes that attack our notions of order, balance, and believability. Hallett Smith, for instance, argues that "from any realistic point of view, the spectacular scenes of *Pericles* are of course utter nonsense" (1529), a position Ruth Nevo seems to echo when she talks about "the dream-like aspects of its representations" (151). Kay Stockholder maintains that it is less a "naturalistic world" than a "highly symbolic world" (18) that we see in *Pericles*, while Constance Jordan views the varied settings as stages of a journey for Pericles "in his acquisition of royal discipline" (344). The critics agree: it is an unrealistic play, but no one posits a relationship between

Figure 5.3 The Cracovia Monster—"*The cause of this misshapen, monster... [is] the detestable sinne of Sodomie*" (Rueff 157–58).

narrative and nature, between the ways that a violently unnatural world seems to require a violently disrupted sense of narrative order and credibility.

If predictablity defines order, then unpredictability (at the heart of ecophobia) is the essence of chaos. *Pericles* dramatizes chaos on the level of narrative while writing nature as unpredictable, inconsistent, and dangerous. It is the enemy. Yet, as always, it is an ambivalent place. The sea, for instance, is at once a place of safety to which Pericles flees when, disgusted and in mortal danger, he discerns the true character of the King and daughter's relationship. It is also a violent monster that in act 2 eats up a ship full of people but inexplicably spares Pericles. It eats up his wife later in the play and spits her back out. Still, the sea seems the only logical place for Pericles to go. Indeed, Harry Berger, Jr., seems correct in arguing that in Shakespeare,

> The green world is…ambiguous: its usefulness and dangers arise from the same source. In its positive aspects it provides a temporary haven for recreation or clarification, experiment or relief; in its negative aspects it projects the urge of the paralyzed will to give up, escape, work magic, abolish time and flux and the intrusive reality of other minds. (36)

Certainly, we see in *Pericles* what we see repeatedly in Shakespeare— namely, that much of the discursive utility of the natural world resides in its not being pinned down: keeping it ambiguous means keeping it perpetually useful. It is a slave that will do any job, whether it is in support of homophobic discourses or as a resource for furnishing repose from corrupt civilization.

But if nature and narrative are violently disrupted in this play, the cannibalism motif, introduced through the eater of mother's flesh, runs the breadth of the play and is always in some ways linked with the topics of sex, appetite, and insatiable passion. We learn in the fifth act that Pericles laps up the words Marina speaks. She "starves the ear she feeds, and makes them hungry, / The more she gives them speech" (112–13). Thaisa herself earlier expresses her desire for Pericles in cannibalistic terms:

> By Juno, that is queen of marriage,
> All viands that I eat do seem unsavory,
> Wishing him my meat. (2.3.30–32)

What is less obvious and more surprising is how these topics are spatialized through competing geographies that supercede one another as they vie for ascendancy in the race to become raw material for Pericles and his plot, and it is here that cannibalism and incest again take form as ecocritical issues.

The first place Pericles lands after fleeing the freaky father and daughter is Tharsus, where the people are complaining about their woeful situation, about how Tharsus, once rich, has fallen on bad times. It is an insatiable appetite that has brought Tharsus down:

> These mouths who but of late earth, sea, and air
> Were all too little to content and please,
> Although they gave their creatures in abundance,
> As houses are defil'd for want of use,
> They are now starved for want of exercise. (1.4.34–38)

It is in this context that the topic of cannibalism returns with

> Those mothers who, to nousle up their babes
> Thought naught too curious, are ready now
> To eat those little darlings whom they lov'd.
> So sharp are hunger's teeth (ll.42–45)

Cannibalism is also the topic of some Fishermen Pericles comes upon after leaving Tharsus. One Fisherman, comparing the marine economy to the social economy of humans, says

> the great ones eat
> up the little ones. I can compare our rich misers to
> nothing so fitly as to a whale: 'a plays and tumbles,
> driving the poor fry before him, and at last devour
> them all at a mouthful. Such whales have I heard on
> a' th' land, who never leave gaping till they swallow'd
> the whole parish, church, steeple, bells, and all. (2.1.28–34)

Cannibalism lacks affirmative associations in the play but always has spatial implications, and its presence in a given geography keeps Pericles on the move.

The fact that deformities and monstrosities hobble and twitch all over the Renaissance stage is crucial to our understanding of the ways that metaphors and classifications work, both in the prodigy/ monster genre and in texts not expressly dedicated to offering definitions. Deformities and monsters, belonging to neither "culture" nor "nature," inhabit and bound sites of disorder.

Ambroise Paré explains that "monstres sont choses qui apparoissent outre le cours de Nature" (monsters appear outside the course of Nature), while "prodiges ... sont choses qui viennent du tout contre Nature" (prodigies are completely against Nature) (3). He does not, however, say *where* either are. Bruno Latour argues that the modern conception of hybrids and monsters is "as a mixture of two pure forms" (78), which, by following his discussion of subject and object, we can eventually trace as being culture and nature. Effectively, this, as in Paré, puts the monster outside of and in conflict with nature. Latour's understanding of the early modern imagination, like Paré's theory, writes nature as an enforcer of strictly defined aesthetic and moral parameters. Within such parameters, nature either rejects certain beings and behaviors or endorses them. Most definitions of monstrosity, in fact, seem to write nature in a similar fashion. Donna Haraway, for instance, maintains that "monsters have always defined the limits of community in Western imaginations" (180) and that "nature and culture are reworked" by monsters and cyborgs (151), while Keith Thomas argues "that monstrous births caused such horror [in the early modern period in part because]...they threatened the firm dividing-line between men and animals" (39). Though often putatively pitted against nature, monsters are the embodiment of the broken boundaries, confusion, and chaos that defines ecophobic conceptions of nature. Indeed, as Jeffrey Jerome Cohen explains, the monster is "a kind of a third term that problematizes the clash of extremes" (Cohen 6), a border-crosser whose "very existence is a rebuke to boundary and enclosure" (7).[18] Tolerable nowhere, abjected, monsters are unassimilable: Julia Kristeva argues that "the unassimilable alien, the monster...strays on the territories of *animal*" (12–13). "Strays" is a good word, since, as "aberrations in the natural order" (Park and Daston 22), they cannot reside *within* those territories.

Important not only in the conceptual mapping of nature and culture, the problematical category of monstrosity is also instrumental in imperialist narratives whose goal is to write racial difference clearly, definitively, and persuasively: Katherine Park and Lorraine Daston explain that "monstrous races—men with a single giant foot, or huge ears, or their faces on their chests—had played a part in the descriptions of Africa and Asia since antiquity and still figured in Renaissance cosmography" (37).[19]

Staging cultural supremacy for the early moderns means, to a large degree, articulating clear relationships between hostile people and hostile lands. It means staging ecophobia. Chris Tiffin and Alan Lawson's argument that colonialist discourse "alternately fetishized and feared its Others—both race and place" (5)—is pertinent to drama of this period, where both race and place are subject to a politics of domination, demonization, and exploitation.

6

DISGUST, METAPHOR, WOMEN:
ECOPHOBIC CONFLUENCES

> To feel disgust is human and humanizing
> —William Ian Miller 11

When asked to compare England and Fez in Thomas Heywood's *The Fair Maid of the West, Part 2*, Clem explains to the black Queen Tota that she holds England "to be the cleanlier" (1.1.72–73). Claiming that the British (the whites, in other words) "never sit down with such foul hands and faces" (1.1.75–76) as the blacks of Fez, Clem rehearses a familiar text, where the polluting person, and his or her physical space, is unacceptable, disgusting. Indeed, as Mary Douglas has argued in her monumental *Purity and Danger*, "The polluting person is always in the wrong" (114). The contours of this wrongness both define the boundaries of the human and imply limits to ethical considerability of all that lies beyond those boundaries. Filth, of some kind or another, becomes a precondition for difference and exclusion. As Stephen Greenblatt argues in "Filthy Rites," "The very conception that a culture is alien rests upon the perceived difference of that culture from one's own behavioural codes, and it is precisely at the points of perceived difference that the individual is conditioned, as a founding principle of personal and group identity, to experience disgust" ("Filthy" 61).[1] Disgust, then, designates difference, but the patterning of disgust, because it is constitutional in determining ethics, is also involved in configuring ecophobia, the writing of hostile geographies, Horace's *terras domibus negata* (*Odes*, 1.22.22).

ROT

In *The Hydra's Tale*, Robert Wilson identifies a "thin drizzle of filth that rains constantly upon the fictional world of *Hamlet*" (10–11)—a drizzle we might be more inclined to see as a torrential downpouring of rot and decay—and argues that the text repeatedly, though not explicitly, imagines disgust. It is disgust that more often than not grows out of rotten environments. We know the famous explanation that "something is rotten in the state of Denmark" (1.4.90), and while the word "rotten" is metaphorical here, suggesting perhaps more about moral turpitude than about green issues, the play consistently conceptualizes the disgusting *as* nature, which is essentially disordered in this text. For instance, Hamlet's description of his world as "an unweeded garden / That grows to seed; things rank and gross in nature / Possess it merely" (1.2.135–37).[2] In a play that sees human disorder in environmental terms, permanence is ugly and "brevity is the soul of wit" (2.2.90) and beauty. Excess is ugly. The "violet in the youth of primy nature / Forward, not permanent, sweet, not lasting" (1.3.7–8) is acceptable, good, and beautiful; gardens rankly overgrown in this play poison "the whole ear of Denmark" (1.5.36), and the "fat weed / That roots itself in ease" (1.5.32–33) in this garden is Claudius, whose "offence is rank, it smells to heaven" (3.3.36). Even the sweet "rose of May" (4.5.157), Ophelia, becomes a site/ sight of floral excess, bedecked with "fantastic garlands . . . / Of crowflowers, nettles, daisies, and long purples" (4.7.167–68). Ophelia, "a document in madness" (4.5.176), is Other, and environmental excess in *Hamlet* is a finger pointing directly at this variety of Otherness.

The metaphors Hamlet uses are very telling. Whenever he talks about difference, his thoughts eventually devolve upon some form of rot. For instance, evil resides in excess, and people are bad only

> By their o'ergrowth of some complexion, / . . . /
> Or by some habit, that too much o'erleavens
> The form o f p lausive m anners . . . these m en / . . . /
> Shall in the general censure take corruption
> From that particular fault. (1.4.27–36)

The problem is *not* "one defect" or "particular fault," since nobody is perfect; the problem is the "o'ergrowth" of such a "complexion." Excess (and eventually rot), then, is the problem, and it is defined with naturalistic imagery. For Hamlet, the social world is rotten to the core, and at a time in history "when the universal belief in analogy and correspondence made it normal to discern in the animal world a mirror image of human social and political organization" (Thomas 61), it is hardly surprising to hear Hamlet ecophobically condemn the natural world as "but a foul and pestilent congregation of vapours" (2.2.302–3).[3] It is dirty and disgusting, like "the rank sweat of an enseamed bed, / Stew'd in corruption" (3.4.92–93).[4] Hamlet is obsessed with rot, with "rank corruption, mining all within, [that] / Infects unseen" (3.4.150–51), with "the sun breed[ing] maggots in a dead dog" (2.2.181), and such issues. This is a man whose strong concerns with purifying his social world results in a discursive putrefying of the natural world. His world is, metaphorically speaking, filthy and rotting, polluted beyond repair.

What makes rot of such concern to theories about ecophobia is—among other things—its imagined unpredictability, its willy-nilly transgressions and blurring of borders, and its perceived alliance with an antagonistic nature. Moreover, as one might expect, this rhizome of ecophobia reaches into matters of gender. If Leonard Tennenhouse is correct to urge that early modern tragedy "defines the female body as a source of pollution...[and that] any sign of permeability automatically endangers the community" (117–18), then the female rape victim becomes a site of pollution (as Ophelia's tousled hair perhaps signifies), and the woman with her own sexuality is also a site of pollution (and a threat to the patriarchal hegemony). Primarily the transgression of culturally significant boundaries, bodily orifices being one such set of boundaries, pollution becomes matter of both gender and environment. Texts represent women as sites of pollution perhaps, as Linda Woodbridge explains, because "women have more orifices than men to start with, which may be why the female body offers the more frequent image of society endangered" (*Scythe* 52). Yet, there is obviously much more going on in the gendering of pollution than orificial politics.

Definitions of pollution are a tricky business, laden with ideology. Douglas defines the polluting person as one who crosses clearly defined "lines of structure, cosmic or social" and adds that "he has developed some wrong condition or simply crossed some line which should not have been crossed and this displacement unleashes danger for someone" (*Purity* 114). The use of the male pronoun here is unfortunate, since more women are perceived or constructed as sources of pollution than men in early modern drama.[5] The tradition that identifies women as a source of pollution is concerned not merely with what goes in but also with what comes out of the body. Thus, women who speak out of order become sites of pollution as do menstruating women.[6]

Since there are, as Karen Warren (among many others) has convincingly explained, "important connections between how one treats women, people of color, and the underclass on one hand and how one treats the nonhuman natural environment on the other" ("Introduction" xi), it seems partial at best to conduct ecocritical investigations outside of feminist frameworks, especially when ecocriticism prides itself on making connections. Again, though, terminological questions arise. Noel Sturgeon's question about "what's in a name"[7] remains germane, as does her suggestion for a plurality of ecofeminisms. Nevertheless, one is tempted to agree that the very term "ecofeminism," whether plural or singular, might "only be transiently useful within our history" (Sturgeon 168), though I would hesitate to suggest that we have exhausted its usefulness. Basically, all ecocriticism must, by its nature, be feminist.

One of the consequences of drawing a distinction between ecocriticism and ecofeminism is that we immediately seem to establish an agonistic discourse that sets ecofeminism and ecocriticism against each other as competing voices, perhaps even as a sort of gender war writ small in the rarefied airs of competing theoretical discourses. It is not an argument that I particularly want to develop, since it is far less productive than building on the strengths of each approach, looking at ways that they complement each other, and working toward defining more fully what each approach envisions. Another problem is that differentiating between ecofeminism and ecocriticism lands us in a bit of a Catch-22: in choosing ecofeminist approaches, we privilege the social; in choosing ecocritical approaches, we subordinate feminism and make it a topic for inclusion rather than a primary topic. Nevertheless, there remain unexamined differences between the two approaches.

When Ynestra King argues that "in ecofeminism, nature is the central category of analysis" ("Healing" 117), she is surely mistaken. Mary Mellor explains that "although ecofeminists may differ in their focus, sex/gender differences are *at the centre* of their analysis" (69; emphasis added). Most ecofeminist scholars agree on the primacy of sex/gender differences over nature as "the central category of analysis." It is more the case that nature is included in the discussion. Despite prioritizing nature in ecofeminism, King seems to agree with this position when she argues that "ecofeminist movement politics and culture must show the connection between all forms of domination, including the domination of nonhuman nature" ("Toward" 119; emphasis added)—including, but not beginning with it. As Greta Gaard and Patrick Murphy observe, this inclusionary view has been "generally embraced as a sound orientation" ("Introduction" 3).

So even though "eco" comes first in both terms, in "ecofeminism" it is the second part of the term that has ontological priority. This emphasis means that ecofeminism is first a social theory, a human-centered approach; ecocriticism tries to be something else, to move away from homocentric models, to put the puzzle of which humans are part before the piece.

Although both ecocriticism and ecofeminism, of various shades, have looked at and added to our knowledge about assorted meanings of particular metaphors linking women and the environment, the social and the natural, and so on, more work is needed that challenges "metaphor," that examines the conceptual maneuvers performed by "metaphor," that questions how we might measure the quality and quantity of change "metaphor" can cause in how we perceive and act in the world, and that analyzes in materialist terms how coercion (as well as meaning) resides in form. Because so much of nature in literature is deeply metaphorical, these seem to be crucial questions ecocriticism needs to address.

METAPHOR AND ECOPHOBIA

Whether we are critiquing ecophobia, misogyny, racism, homophobia, classism, or any other oppressive way of thinking and acting, metaphor is clearly important. But precisely what counts as metaphor? And, since everything ever spoken, written, or thought is framed (or based, depending on which metaphor we want to use) in metaphorical terms, does not the distinction between figurative and literal become spurious? Certainly an unproblematized use of the term assumes that "reality" is, to use Andrew Ortony's words, "literally describable" (1). Numerous rhetoricians, psychologists, and linguists have argued a similar position, but even among the specialists, there is no real consensus obviating the category of metaphor—largely, we might surmise, because there is great functional value in keeping alive a thing such as the figurative/literal binary.[8]

Whether or not we abandon the binary opposition, though, it seems imperative to accept the view David Rumelhart holds (that Jonathan Culler, Paul de Man, I. A. Richards and many others from different areas of specialization have also held)—namely, that "metaphor in language is absolutely ubiquitous and the existence of nonmetaphorical language is questionable" (81).[9] One of the premises Rumelhart works from is that "the distinction between literal and metaphorical language is rarely, if ever, reflected in a qualitative change in the psychological processes involved in the processing of that language" (72). Accepting the ubiquity of metaphor in language and rejecting the figurative/literal binary, however, presents a whole new set of problems—not the least of which is in having to deal with (or name), for instance, a statement "that appears to be perversely asserting something that is plainly known not to be" (Black 21). We know, as did the Elizabethans, that the moon, though it appears to have a face, cannot be "envious"; yet Romeo calls on the sun to "kill the envious moon" (2.2.4).[10] But it is more than just a definitional problem[11] that we are facing here: we are also confronted with the difficulty of measuring the quality and quantity of change that such language causes in our perceptions of the world. It is this change that I see as the central issue "metaphor" presents for ecocriticism, for it is here that we can best see the ideological workings of texts that habitually link natural and social worlds—and more often than not in ways that are (1) disempowering to marginalized groups, (2) ecophobic, and (3) powerfully coercive because of their freshness.

One of the central arguments Max Black makes about metaphor is that "a metaphorical statement can sometimes generate new knowledge and insight by *changing* relationships between the things designated" (35). It is an argument that George Lakoff and Mark Johnson take one step further when they conclude that metaphors often provide "the only ways to perceive and experience much of the world" (239). For Lakoff and Johnson, metaphoric thinking is the *basis* for all knowledge, understanding, and, in particular, language.[12] It determines the way that we organize our material realities, the material things that we use, the worlds we create and destroy, and so on.

Literature written in English is rife with metaphors that link women with the natural environment, and useful feminist, empirical, and even thematic commentary has been offered.[13] There is little, however, in terms of a theory that analyzes the material effects of metaphor. Except for some broad statements about early modern correspondence and analogy theory, there is surprisingly little work that actually theorizes *how* metaphor works in producing and perpetuating material practices that hurt people (and the environment). Of all that has been written on Shakespeare, for instance, there is only one full-length monograph that attempts a serious study of metaphor: Ann and John Thompson's *Shakespeare: Meaning and Metaphor*. The book begins with the Lakoff-Johnson model to look at time in *Troilus and Cressida*, moves to semantic field theory to discuss animals in *King Lear*, jumps to metaphorical theories of Group M to analyze metaphors of the human body in *Hamlet*, on to analogy theory from J. S. Ross to discuss "Sonnet 63," and finishes with some test cases for Donald Davidson's theories about printing metaphors. As a whole, the book lacks cohesion in its eclectic hop from one approach to another. Not being concerned with environmental issues on any level (and predating ecocriticism by about a decade), it unsurprisingly makes no kind of committed comment about the natural world—the remarkably apolitical chapter on *Lear* being a case in point. While the book bravely takes Shakespeare somewhere new, it does not take us into the poorly mapped terrain of metaphor theory, the soil out of which social praxis (and metaphor generation itself) grows.

Neither ecophobia nor misogyny, racism nor homophobia, social hierarchy nor ethnocentrism stand unsupported by the ever-present effects of metaphors. The most direct way into this complex issue is through an analysis of what metaphor does—and how.

Archeological theorist Christopher Tilley asserts the brilliant and accessible thesis that the natural world often "structures an entire series of values and attitudes that pervades the manner in which...people live" (50) and, citing a variety of theorists, explains how "metaphorical talk often presupposes and reinforces an intimacy between speaker and listener" (9).[14] These two observations are important if we are discussing Shakespeare through an ecocritical framework because they allow us to question what ways the audience, both in contemporary and Elizabethan times, are made complicit, forced into ways of thinking, both about figures of otherness and about the environment. We must be very careful, though, not to fantasize or romanticize the past. It is easy enough to make this mistake when we think on the fact that in the time of Shakespeare, the environment, globally, was in much better shape than it is now. Locally, however, it was a different story in many metropolitan centers of early modern Europe.[15] Lacking the popular environmental ethics we have (or profess to have) today, urban centers were often heavily polluted and deadly. The claim that both Keith Thomas and Carolyn Merchant make about our having lost some better way of thinking about the natural environment, therefore, seems misled.

Merchant, for instance, begins her book talking about "the world we have lost" (1), while Thomas laments the shattering of analogical assumptions: he explains that

> the naturalists...shattered the assumptions of the past. In place of a natural world redolent with human analogy and symbolic meaning, and sensitive to man's behaviour, [the early moderns] constructed a detached scene to be viewed and studied by the observer from the outside, as if by peering through a window, in the secure knowledge that the objects of contemplation inhabited a separate realm, offering no omens or signs, without human meaning or significance. (89)

Neither argument acknowledges that the good ol'days, in fact, were not so good, conceptually—never mind materially. An analogical anthropocentric conception of the cosmos is *inevitably* bound to produce no good. The change in thinking about the environment from organic to mechanistic world views is a linear development and not a paradigm shift or paradigmatic hop, is less of the discontinuity Michel Foucault describes in *Les Mots et les choses*, more of a progress from than a movement outside of organic thinking,[16] and the anthropocentrism, the utilitarianism, and the power dynamic remain essentially intact, though amplified in some ways. A baby does not become a different species when he or she grows up; neither does organicism of the sort Thomas and Merchant describe.[17]

Partly, of course, this development, this change in the metaphors through which people understood and organized the natural world, has to do with the increasingly global view of things that various explorations afforded, and to at least some degree the changes in views toward the natural world are changes in ideas about space and about the little island of England.

METAPHOR, WOMEN, ENVIRONMENT: A CASE STUDY OF *THE WINTER'S TALE*

Certainly space (and how it is conceptualized) in a play such as *The Winter's Tale* is very important, not only for the choppy plot but, more importantly, because it determines the structure of the lived experiences of the people in those imagined spaces.

The trouble in *The Winter's Tale* begins in the domestic space of King Leontes of Sicilia and his wife Queen Hermione when their guest King Polixenes of Bohemia is preparing to leave. The polite entreaties of Leontes for Polixenes to stay longer fail, but Hermione manages to prevail upon the Bohemian King. Leontes sees something amiss in all of this and, wrongly believing his wife to be flirting with Polixenes, asks one of the lords of his court, Camillo, to kill Polixenes. Refusing, Camillo flees with Polixenes. More demented than rational at this point, Leontes has his pregnant wife imprisoned, and when Perdita (who is his daughter) is born, he orders her to be abandoned in a desert, since he believes her to be the child of Polixenes. While his wife is in prison, his son Mamillius, meanwhile, dies of grief. His daughter, his son, his wife, and his friend are each withdrawn from the space he seeks to control.

Seeking answers about whether he has behaved properly, he sends two lords of his court out to the oracle at Apollo's temple at Delphos for answers, who responds that Hermione is innocent and that Leontes is a tyrant. He refuses to heed them until he hears about his son's death, and shortly after is informed that his wife is also dead. She is not, and for sixteen years, she hides at the house of her friend Paulina. Paulina's husband, Antigonus, is one of the people commissioned to take the baby girl to the desert, and, after doing so, he is ripped apart by a bear that appears suddenly from nowhere. Perdita grows up in Bohemia thinking that she is lowly country folk and is courted when she is sixteen by Polixenes's son Florizel. The main obstacle they face is their imagined class difference. They eventually learn the truth, return to Sicilia, and marry with the consent of both Leontes and Polixenes, now reconciled. Paulina takes everyone to her house in celebration of the wedding and presents a "statue" of Hermione, which turns out to be no statue at all, but Hermione herself beguiling art that would "beguile Nature" (5.2.99) with the reality of her being.

Throughout the play, there is a preponderance of metaphors linking women and the natural world. These offer relatively easy access to some fairly complex issues—relatively easy because approaching the text through these metaphors resembles (and can too easily swing into) a formalist thematic groove (which students tend to prefer because it is simpler than, say, materialist criticism). The representations of women and the environment clearly articulate values about patriarchal power that the text carries. Both the environment and women are characterized in ideologically highly charged terms. The environment and women are often *either* good *or* bad in Shakespeare: in *The Winter's Tale*, the environment is a vicious space of bears and wolves, or else a beautiful place of fertility and abundance; women are liars, shrews, and lechers all, or else they are chaste, guiltless, or otherwise guileless. There is no ambiguity in this play. Paulina is a good woman, as is Hermione,[18] but the spectator (constructed with all of the insecurities of a man like Leontes) is dragged along and made complicit in the testing of these women. Justifiably or not, the audience may wonder about Hermione and about whether Leontes has just cause in his worries. This raises several questions that are difficult to answer but useful in the classroom: Do the men and women in the class have the same thoughts about Hermione? Where do these responses come from? What ideological positions do these responses to Hermione support?

These are important starting points, but they are only starting points. For any woman attacked by a man, ideas that women and the natural world are metaphorically linked are unimportant compared with what thoroughgoing analyses of the material implications

of such linking can give us. Simple formalist notations about image
clusters and metaphorical correlations and what they prove or dis-
prove about Shakespeare's personal agenda or his deftness at weaving
together coherent patterns can only give us so much information.
Bound up with metaphor in *The Winter's Tale* are important questions about language and how it organizes materials in the play. The
person who controls (or seeks to control) language and strategies of
representation in the play is Leontes, from whose frustrated desire for
control the main action of the play springs.

The importance of language in *The Winter's Tale* has been succinctly noted by David Laird, who argues that the main problem
for Leontes is in keeping a sense of order and goodness and that it
is a linguistic problem: "To control language, to exercise the power
to name, categorize, and classify is an essential weapon in the arsenal" of things Leontes uses to control his world; so, when Leontes
thinks that Hermione uses "a discourse where meanings are multiple, ambiguous," we see an unstable patriarchal discourse striving to
consolidate itself yet beleaguered by disruptions of order, transgressions, instances of pollution, and so on that are in some ways radically
subversive and in other ways are simply versions of licensed release.[19]
Laird goes on to say that "Hermione speaks a discursive skepticism
that measures the distance between words and things" (27); unfortunately, he does not explore *how* this relationship between words and
things functions in the objectification of Hermione, how, in other
words, words thingify her.[20] She is a palpable material presence in the
text, yet the text vigorously excludes her from much of the material
action of the drama, the male action that determines her material
fate. Made passive, excluded, and ossified, Hermione may be, as Laird
implies, "singularly daring" (30), but she suffers singularly in a way
that singularly daring men in Shakespeare do not.

This is a time when it is "only natural" for women to be excluded
from the material action of cultural and public life.[21] There are, of
course, noteworthy exceptions to this exclusion, the greatest, perhaps, being the Queen herself. Yet, even she participates in both an
excluding of women from public life and in an emulation of maleness.
Nowhere is this clearer than in her speech to the troops at Tilbury on
the approach of the Spanish Armada: "I know I have the body of a
weak and feeble woman," she says, "but I have the heart and stomach
of a king." It is a startling self-effacement, yet one that responds to
the social organization surrounding Elizabeth.

In *The Winter's Tale*, the language (and, in particular, the metaphors) force us to participate in the organization and experience of

the play's social and material world. As with Prospero in *The Tempest*, Leontes is a man whose power resides in language,[22] a man who controls women, a man who treats women and the environment as passive objects that, ideally, lack their own volition and voice. But while *The Winter's Tale* fails to challenge ecophobia,[23] it does manage, though inadvertently, to challenge gynophobia. It is a mistake to read for anything more than inadvertence here. The play comes down with forceful conservatism against any kind of challenge to gynophobia that it may have set up when it asserts that the reactions Leontes has to Hermione and Paulina are unbased, that there really are no evil women in the play, that Hermione is evil only in the mind of Leontes, and that Paulina's open revolt against constituted authority is for a higher moral good than that which the crown pretends to represent. The phobic reactions Leontes has toward Hermione and Paulina are highlighted as anomalies, strange and condemnable aspects of this lion of a man, this man who is all the more outside the pale of human intercourse by his bestializing name. The play firmly maintains that the misogyny Leontes feels cannot be rationalized[24]— which, of course, begs the question: would his misogyny be justifiable if Hermione and Paulina *were* evil? This lion seems a bit barmy, and for this barminess, the audience is unforgiving.

Yet, it is not so much his misogyny that we condemn; it is his lack of justification for it, and we are drawn into accepting this whole weird equation. The audience is written into a position of complicity, of an acceptance of the anomalousness of Leontes, which effectively diverts our attention from what seems to me the more important question: the issue of patriarchally sanctioned misogyny encoded in our acceptance of our lion's anomalousness. In accepting that Hermione and Paulina are not evil, we tacitly accept that things would be different if they were evil, that the behavior of Leontes might be more acceptable.

Critical tradition has read *The Winter's Tale* as political, religious, and autobiographical allegory; as fantasy; as geographically improbable; as the work of someone other than Shakespeare; as realism par excellence; as the literature of escape; as a sophisticated vegetation myth; as boring; as a falling off; as a structural, thematic, or philosophical experiment; as a general failure; as a perfect example of symbolic technique; and so on. There have been reams written on the bear that runs off with Antigonus; discussions about the tension between art and nature in the play are everywhere; and there have certainly been enough analyses of the role and function of natural imagery in the play. Nowhere, however, do we find discussions about the spatial dimensions of the play's patriarchy, of how the different geographies work in maintaining certain kinds of social relations, or of how the natural environment works in all of this.

Banishing his family and friends from the space that he seeks to control, Leontes creates a sort of psychological no-fly zone, a never-never land where he can continue to act with all of the puerility that eternal childhood would mean for a boy. And it is clear that "to be a boy eternal" (1.2.65) is a shared fantasy of Polixenes and Leontes. But if it is fantasy for the former, it is something more for the latter, and this probably explains why he has such a fit at the mere thought that his boyhood friend *could* conceivably have adult urges. Leontes keeps his domestic space just the way he wants it. The metaphors that structure the play's understanding of the spaces to which the unwanted are consigned offer an image of a sentient and virtually human natural world, one that punishes and rewards in the same way that people do. It kills the villain with one of its bears but spares the innocent baby left in its stormy midst (and even yields up gold for the child's future!). Spatially, the deserts of Bohemia may be a world apart from the safe never-never land Leontes calls home, but conceptually, they are one and the same, more extensions of people organized by their passions than spaces running on their own rhythms and principles. Certainly the green fantasy where Perdita grows up with her shepherd stepfather is a stark contrast to the sterile and life-denying court of Leontes. But Perdita, as her name implies, is lost, and the fold to which she ultimately returns does not have sheep in it: she ultimately returns to Sicilia, and the play's pastoral interlude remains just an interlude. Nature is a place people go to visit, but living there permanently is not something that people of class do. It is logical that such is the psychology in the play (indeed in the early modern period, if not today) since the space of nature is the space of commodities.

If nature is made to resemble people in the psychology that is ascribed to it, we do not see pain or suffering similarly ascribed to nature. When it suits the play to anthropomorphize, it anthropomorphizes. Nature may be manipulated and exploited (cross-breeding is the chief example in this play), but it does not suffer or complain. Not so for people who are manipulated and exploited. People *do* suffer in this romance. Hardly a romance for Mamillius: he dies. Hardly a romance for Hermione: she has half her life taken away and for sixteen years has no daughter. And why? While we are, for the most part, spared the kind of disaster we see in other plays[25]—perhaps because the play is generically confused, beginning in high tragic style and switching abruptly to comic mode with the sudden appearance of the

bear in act 3, scene 3—what we do get in the jiggling of classifica-
tory schemata, the blurring of "natural" and "unnatural," of culturally acceptable and unacceptable, of fair issues and monstrous ones,
is the staging of disgust—one of the great theoretical unattendeds in
Shakespeare.

For a play that foregrounds the pastoral tradition so heavily, that
stresses so insistently a relationship between nature and art, that is so
deeply rooted at so many levels in conceptual dividedness, there is a
remarkable wealth of things we can say about the pollution metaphor
that runs throughout the play.[26] The comments Leontes makes about
"mingling bloods" (1.2.109), even if he is wrong about Polixenes
and Hermione, are unchallenged in the play, but the xenophobia
toward mingling bloods does get challenged more substantially in
this play than in any other piece of early modern literature, culminating in the debate between Perdita and Polixines on cross-breeding
(4.4.79–108).

Cross-breeding, by this time in history, was an established practice of the nursery business, though it was not without its anxieties.
These anxieties have to do with threats to order, with confusion,
with mixing of categories, with monstrosity, with bestiality, and so
on. Cross-breeding is a practice important to the demands of the
early modern economy,[27] and it has numerous implications, both
in the play and in early modern cultures in general. It is an important question regarding the protagonist couple, Perdita and Florizel,
who, to all appearances, are mismatched: Perdita, ostensibly a country lass; Florizel, a prince. The whole section on what Perdita calls
"Nature's bastards" (4.4.89–96) smacks heavily of allegory: if there
is any initial doubt about whom the gentler scion or the wildest stock
might refer to, it is dispelled a moment later by Perdita when she
talks about Florizel breeding by her (4.4.103). In an instant, she has
co-located women with breeding animals and fertile flora. Yet, this
is the same woman who sees cross-breeding as a diluting of nature,
a hybridization and infection of natural processes: "I care not to
get slips from them [cross-bred things]" (4.4.83–84), she insists,
because, she thinks that selective breeding "shares / With great creating Nature" (4.4.87–88). The argument that Polixenes makes is
that Perdita (of ostensibly wild stock) and Florizel (a gentler scion)
can cross-breed profitably and without fear of the kind of pollution
Perdita seems to imagine. Polixenes has argued that in a material
sense cross-breeding, rather than polluting to nature, is, in fact, natural: it uses natural materials. Polluted nature is a loathsome thing
to be avoided at all costs.

Cross-breeding, nevertheless, a form of pollution in the text as in the larger culture of which the text is a part, disrupts and blurs classification systems, and people obviously suffer when there is this kind of jiggling and the disgust and revulsion it produces and embodies. Disgust is central, as Richard Twine has recently argued, in "the emotional repertoire of the historical emergence of specific exclusionary and hierarchical deployments of the 'human'" (Twine 2).[28] Disgust is constitutional in defining the "human" and is both the effect and agent of imagined unpredictability, of willy-nilly transgressions and blurring of borders, and it ties together a host of theoretically related issues. Woven into metaphors, discourses of rot and pollution show clear links between ecophobia, misogyny, and the ways that metaphor itself works.

7

STAGING EXOTICA AND ECOPHOBIA

Tomaso Garzoni begins *Hospitall of Incurable Fooles* (1600) with a startling description that draws together images of monstrosity and sheer ugliness with the general category of "folly," at once defining madness as female, monstrous, and polluted, while representing a composed mind as male, pure, and, oddly, passive. It is a strange inversion of metaphor, where the colonizer is female and the colonized male: "Folly…is she, who being spred and dispersed, ouer all prouinces and countries in the worlde; sorely vexeth mortall men, and holdeth in subiection under her tyrannicall empire, an infinite number of people and men" (2). This kind of metaphor, unusual in that it implicitly associates men with the land, is consonant with metaphors that depict women as polluted and polluting. As we have seen in previous chapters, discourses that produce difference in terms of commodifiable attributes configure corporeal diffraction as the normative ideal for the body of the Other; yet theorizing this corporeal diffraction does not address the ways in which discourses of less-apparent corporeal significance (specifically discourses of madness) spatialize, transcode, and commodify bodies. The transcodings between the Other and the bestial in discourses of madness, as this chapter shows, is at once speciesist and ecophobic, and there is a generalized environmental loathing implied in the exoticization of early modern Others.

Madness, like sodomy, is a slippery term, but whatever it is taken to signify is being reconstituted in the Renaissance. Madness, of course, is one of the most commonly depicted deformities about which the literature of the time expresses vigorous pruning and reformation efforts. Madness, though also represented through metaphors of monstrosity and as frequently described as being manifest in discursive or material action, is understood more in ontological than necessarily behavioral terms, as a state of being that reveals itself in outward signs, often behavioral but always material—either as deformity or action. In the Middle Ages, it "marked the inter-section of the human and the transcendent" (Neely, "Documents" 317). At this time, demoniacal possession was a holistic experience, deforming body and mind: as Edgar Allison Peers reports, "It was made to account not only for mental disease but for all kinds of physical deformations and imperfections, whether occurring alone, or, as is often the case, accompanying idiocy" (8). The tradition res-onates but is waning in the Renaissance. Exorcisms to cure madness are staged in Shakespeare—for instance, in *Twelfth Night*[1]—and the language of possession, as Duncan Salkeld notes, is also found in such plays as *King Lear*[2] and *The Comedy of Errors*, but "in each case what [is] referred to in such language are cases of spurious possession. The spiritual potency of the terms has been 'emptied out', to use Greenblatt's phrase" (Salkeld 25—see also Greenblatt, "Exorcists" 119). But if the spiritual connection is subsiding, the corporeal significance is taking prominence. And it is taking promi-nence in very specific ways.

The corporeal norm onto which madness is written as material deformity is gendered, sexualized, classed, and raced. The deform-ing of this norm tacitly accepts and employs an ecophobic ethic that needs to be addressed if we are to understand how environmentalism, far from being removed "from a politics of personal liberation and empowerment" (Kerridge 6),[3] has a vital role to play in liberation movements.

In other words, because discourses of monstrosity so often use images of the natural world as a part of their processes, a politics of personal liberation and empowerment must deconstruct and seek understandings of the implied violence of bestializing metaphors, of pollution metaphors, and of explicitly dehumanizing depictions of deformity. Discourses of monstrosity do more than dehumanize, defile, and deform, though: in producing people who fall outside the community of "the human," they often also declare both species and racial/geographical difference from the presumed species and racial/geographical norms they posit.

When Robert Burton discusses sexual dissidence in terms of for-eignness, the implication is that the normative heterosexual human body is indigenous to the land. The physical land itself is fetishized, becomes a matter of State, of national pride, and of identity, and this patriotism and drawing of boundaries tacitly assumes that everything we need is "abundantly ministered unto us for the preservation of Health at home in our own Fields, Pastures, Rivers, etc" (Culpeper 7, cited in Wear "Making Sense" 128);[4] meanwhile, anything foreign becomes a site and origin of danger, an object of xenophobia and disdain, and a source of pollution.

Burton, addressing his British, literate audience, claims that "this vice [of sodomy] was customary in old times with the Orientals, the Greeks without question, the Italians, Africans, Asiaticks" (651). That "nastiness and abomination" (653) was "something that...the English associate[d] on the whole only with foreigners, not with themselves" (Orgel 39), but the local, while extolled, is not free from pollution. So while foreignness assumes deviance, citizenship does not guarantee "normalcy." Burton's disquisitions represent a homophobic warning directed more or less exclusively at an audience of educated men and boys who can read it,[5] and it is a warning that is remarkably explicit, discussing, among other things, intercourse among men, male orificial pleasures, masturbation, and so on. For Burton, "This tyrant Love rageth with brute beasts" (650): same-sex desire is a sort of beastialism for him, and it is and leads to a form of madness.[6]

What we have here, then, are desires associated with nonhuman animals and the land (other than England). The discourse of sodomy stakes its boundaries clearly: its preference for homogeneity over variety; for segregation over interconnections; for stasis over hybridization and change; for nation over environment; and for people over animals. It is a discourse that is basically antienvironment, cherishing things opposite to what healthy natural ecologies are and do and to what environmentalism seeks. The discourse of sodomy is resolutely ecophobic. To assume that we can address matters of identity outside of an environmental context is a mistake, and we need "to take heed of the ecological impossibility of a purely human or merely human realm within which discussions about identity tend to circulate" (Polk, http://www.queertheory.com/theories/science/deconstructing_ origins.htm), especially since ecological matters are so clearly mobilized in homophobic discourses. For Shakespeare's drama, part of this means revisiting some of the traditional thematic concerns of Shakespearean criticism. Race and monstrosity, for instance, have long been among these, but they have not been connected adequately with environmental matters in the plays.

Hysteria, generally defined in early modern England "as a disease caused by the woman's uterus which floats about her body attacking her...usually [signifying] some aberration in the woman's sexual constitution" (Little 20)—a psychophysical deformity, in other words—is on par with witchcraft, and, in fact, "the symptoms of be-witchment and hysteria are identical" (Neely, "Documents" 320).

Edward Jorden delineates his clinical theory in *A Briefe Discourse of a Disease Called the Suffocation of the Mother* (1603), where he attempts to distinguish between bewitchment and hysteria. He cites two main causes of the disease: internal and external. The "internall causes may be anything contained within the bodie, as spirit, blood, humors, excrements, &c" (F3[V]).[7] Jorden identifies the primary external causes as "meate and drinke" (G2[R]). Whatever the imagined causes for this imagined madness, gender is the sine qua non of deformity in these discourses about hysteria and witchcraft. Such deformity is an environmental issue not only because of the association of women (as a general category) with the natural world, but also because of the many links imagined specifically between witches and the natural environment.

It is more than a simple disruption of order[8] that witches promise: it is a thorough-going threat of confusion of the very boundaries that define the human. We see such confusion, for instance, in the transient corporeality of the witches as they melt into the air, "as breath into wind" (1.3.82) in *Macbeth*. These witches, who "look not like th' inhabitants o'th' earth" (1.3.41), are, as Stallybrass observes, "connected with disorder in nature (not only thunder and lightning but also 'fog and filthy air')" (*"Macbeth"* 195). They are also associated with the undecidable meteorological conditions of "so foul and fair a day" (1.3.38), the likes of which Macbeth has never seen. Further distancing them from proper natural human form was the belief that witches could control the weather, an unnatural power possessed by unnaturally powerful people (among them, Prospero).[9] *Macbeth*'s witches challenge the boundaries of the human through their associations with nature, but their gender-bending also constitutes a deformity from the essentialized notion of Woman: they have beards. As Banquo complains to them, "You should be women, / And yet your beards forbid me to interpret / That you are so" (1.3.45–47).

Oddly, Lady Macbeth actually *seeks* deformity, but it is not so odd if we understand that the individuality implied in deformity frees the subject from the constraints of social, emotional, and physiological conformity. It is, in a sense, potentially very empowering. What Lady Macbeth seeks is "a perversion of her own emotions and bodily functions" (Neely, "Documents" 328) when she rails on spirits in her unsexing speech:

> Come, you spirits
> That tend on mortal thoughts, unsex me here,
> And fill me, from the crown to the toe, top-full

Of direst cruelty! Make thick my blood,[10]
Stop up th' access and passage to remorse,
That no compunctious visitings of nature
Shake my fell purpose, nor keep peace between
Th' effect and it! Come to my woman's breasts,
And take my milk for gall. (1.5.41–48)

Conformity assumes norms, and the assumptions Lady Macbeth makes are, first, that remorse, pity, and kindness are "visitings of nature," and second, that such milk of human kindness is naturally immanent in women and that Macbeth, as a man, is "too full" of this stuff (1.5.17): kindness in a man, like cruelty in a woman, becomes a sign of deformity and goes against nature in this play.

But if there is one thing above all that unmans Macbeth, it is madness. Madness and a muscular heterosexual manhood are largely incompatible with each other in the early modern period. Macbeth is "quite unmanned" (3.4.73) by folly and, though he may complain that his mind is "full of scorpions" (3.2.36), he will always step up when asked, "Are you a man" (3.4.57). The very clear implication here is that real men are not mad; emasculated men often are.

NEW WORLD DREAMS, OLD WORLD NIGHTMARES

With the expansion of the medieval and early modern feudal economies into market-industrial economies, "more and more elements of both the natural environment and human qualities are drawn into the orbit of exchanged things, into the realm of commodities," and this process, as William Leiss, Kline, and Jhally explain, "of converting [natural and human qualities into commodities] constitutes the *very essence* [of the expanding economies]" (273—emphasis added).[11] This has implications for characters as varied as Caliban, Miranda, Shylock, Portia, and Antonio.[12]

The appropriation of the land in *The Tempest* is, as many scholars have noted, inseparable from the appropriation of Caliban. As Chris Tiffin and Alan Lawson explain in the introduction to *De-Scribing Empire: Post-Colonialism and Textuality*, since "only empty spaces can be settled,…the space [of the colonized lands] had to be made empty by ignoring or dehumanizing the inhabitants" (5). In a stroke of pure brilliance, Shakespeare has Caliban very well aware of the strategies of appropriation. Caliban thus laments that Prospero prevented him from raping Miranda: "I had," Caliban explains, "peopled else / This isle with Calibans" (1.2.350–51). The intended rape reveals Caliban's desire to fill the geographical space of the island and the vaginal space of Miranda by force. Rape does not happen without misogyny, any more than colonialist profit-raking from distant lands happens without ecophobia. Both Miranda and the landscape stand as empty spaces waiting to be filled by men.[13] Caliban, of course, is not a man,[14] and he therefore lacks the rights early modern men consider themselves to have to occupy the land and the vagina. The intended rape is a problematic and forcefully Othering form of resistance in the play. The whole venture fails, and Caliban "must obey [Prospero]: his Art is of such pow'r" (1.2.372). Caliban has little power and few of the qualities that entitle him to the kinds of rights the more fully enfranchised men enjoy in the play.

Part of the problem for Caliban is that he is part of the exotica that defines the landscape of the imperialist gaze. The play is a show-case of exotica, of a strange and brave new world, "full of noises / Sounds and sweet airs... [and] a thousand twangling instruments" (3.2.135–37); a land chock-full of wild animals, "toads, beetles, bats" (1.2.340), "the nimble marmazet" (2.2.170), jays, and shellfish; a land where there "is everything advantageous to life" (2.1.50); a land "lush and lusty" (2.1.53); a land that fuels the utopian dreams of Gonzalo, for whom "Nature should bring forth, / Of its own kind, all foison, all abundance, / To feed [the]... people" (2.1.163–65); a land of fairies and monsters, nymphs and goblins; a wild land ready for taking and taming. In many ways, the space that *The Tempest* describes is similar to the "wonder cabinet," which Steven Mullaney describes in *The Place of the Stage* as "a form of collection peculiar to the late Renaissance, characterized primarily by its encyclopedic appetite for the marvellous or the strange" (60–61). It is a space that inspires awed reverence. It is also a space whose Otherness, differ-ence, exoticism, and promise of wealth make it very fertile ground for the seeds of colonialist ambitions and fantasies, and it is a space from which Caliban is characteristic and inseparable. It is at once a space of wonder and danger, of abundance, and of the rot and madness that such abundance implies. It is a space defined by madness, as the opening anchorlessness and craziness of the storm sets up. It is also a space that is totally within Prospero's phallic power (as is Caliban). The environment becomes at times beautiful and serene or hideous and deadly, depending on Prospero's needs.

There is no telling what a man who remakes the environment is capable of, yet, Prospero—though he seems unpredictable—is not imagined in the play as being dangerous, as a force to be feared; the natural world, however, is pictured as dangerous, unpredictable, and

amoral, illustrating an ethics of dislocation from the natural world
rather than one of connection, a mindset of ecophobia rather than
biophilia or ecophilia, a sense of competition between the individual
and the larger opponent of nature rather than a sense of harmony
and mutuality. This is what it means to be living in Prospero's world.
It was a world that arrived on the island and disrupted the natural
harmonies that were there, at least for Caliban.[15] And we know that
Prospero stole this from Caliban, that Prospero, currying Caliban's
trust while drawing on his geographical knowledge, betrayed this
native inhabitant of the "brave new world." Caliban complains

> This island's mine, by Sycorax my mother,
> Which thou tak'st from me. When thou cam'st first,
> Thou strok'st me, and made much of me;...
> ...then I lov'd thee,
> And show'd thee all the qualities o'th'isle,
> The fresh springs, brine-pits, barren place and fertile:
> Curs'd be that I did so! (1.2.331–39)

At another point in the play, Caliban again explains, this time to
Trinculo and Stephano, that Prospero conned him of the island "by
sorcery" (3.2.52), that "by his cunning [Prospero] hath cheated
[Caliban] of the island" (3.2.43–44). Though partially humanized
(at least enough to keep the action afloat), Caliban remains a thing
of difference, such difference as to be a deformity and a monster (all
forty-six uses of the word in this play refer to him). Straddling an
interstitial space between the human and the dangerous natural,
Caliban is subject to various regimes of denial, pruning, and reformation and is finally lopped out of his world and transplanted—a thing
of darkness—into Prospero's world of light.

If xenophobia, ecophobia, and various issues of commodification characterize the old world dreams in the new world, no less
do they characterize Venice, where foreignness also defines the
boundaries both of "the human" and of what makes a man. In *The
Merchant of Venice*, the environmental ethics the play describes is
inextricable from the confluence between xenophobic and (at the
risk of sounding anachronistic) homophobic discourses within the
play.

Shylock and Antonio both lose out to Portia: Shylock is forced
to give up his money and religion, Antonio his Bassanio. Portia/
Balthasar leaves Antonio and Shylock each devoid of what they most
love. Jonathan Goldberg maintains that "[Portia's] power as boy is

directed against and serves to police Antonio and Bassanio and to separate them... [and that] when s/he saves Antonio and defeats Shylock, the two acts form a single gesture—unleashing energies that are racist and homophobic" (*Sodometries* 142). Shylock is downright excluded from, is alien to, Venetian society; Antonio is merely marginalized. Antonio is a figure of difference who lives his apparent same-sex love on the margins of what is human; yet, he *is* human, which is perhaps the reason for the poignancy of the threat he poses.[16] He is in direct competition with Portia for Bassanio's love and, consequently, is a threat to her—if not to the entire heterosexual orthodoxy that the play (with its three marriages) tries so valiantly to stabilize.

The drawing of Shylock as "foreigner" and of Antonio as some kind of sexual Other, a "tainted wether of the flock," that "weakest kind of fruit" (4.1.114 and 115) reveals a confluence not only of anti-Semitic and homophobic discourses,[17] but the allusions to the plant and animal world at least suggest a textual mark of the growing interest in the organized production of untainted agricultural commodities. Nonprocreative sexualities are, in the metaphors of this play, tainted commodities that are dangerous because they threaten the evolving fantasy about an entire cosmos geared toward capitalist production and reproduction. Behaviors that threaten reproduction defy patriarchal thinking about the use and exploitation of natural spaces and commodities—women and land are husbanded, and control of both is the assumed basis of natural order in patriarchal capitalist thinking. Marion Wynne-Davies argues that "control of the womb was paramount to determining a direct patrilineal descent, and when this exercise of power failed and women determined their own sexual appetites regardless of procreation, the social structure was threatened with collapse" (136). Control of nature and of all the monstrosities it houses is a matter of crucial importance to the State. Because nature is coded feminine,[18] the "womb of the earth"—to borrow a phrase from *Hamlet* (1.1.137)—stands as an empty space for men to penetrate and fill. Anything that threatens such penetration also threatens collapse of the social structure and is subject to vigorous disciplinary action.

Discursively self-identified with issues of plant and animal husbandry, Antonio—in whom "commercial capacity has degenerated into effeminacy" (Sinsheimer 92) on account of the uncertain outcome of his penetrations into the physical and economic spaces of mercantile capitalism—waits powerless before Portia, who not only plays the "man's part" better in disguise than Antonio but plays it better out of disguise as well. There is never any question about *her* finances, a fact which, when she is out of disguise, had to have had what Stanley Chojnacki calls "psychological leverage" (128) over the Venetian men around her. In 1568, Edmund Tilney expresses a sentiment that seems in many ways representative of men's fear of a perceived masculinizing effect of money. He claims that "a riche woman, that marieth a poor man, seldome, or never, shake off the pride from hir shoulders. Yea *Menander* sayth, that suche a man hath gotten in steed of a wyfe, a husband, and she of him a wyfe" (Tilney B_2^V). Feminized before the masculinized, cross-dressed Portia, Antonio stands and listens: "Why then thus it is; / You must," Portia tells him, "prepare your bosom for Shylock's knife" (4.1.244–45). And his bosom lies bare.

The feminized male body is held up for view and, possibly, for sacrifice, but certainly for the threat of the penetrating disciplinary knife. Antonio, like Othello, is embodied and disempowered; like Lavinia, he is a body waiting for dismemberment; like the women in the infamous *Regina v. Litchfield* case,[19] he is a body through which patriarchal legal discourses pass. With the fear of death in him, Antonio beseeches Bassanio to "say how I lov'd you" (4.1.275), but when Antonio gets a reprieve from his transvestite opponent, Bassanio may hold his tongue. The love that dare not speak its name now *need* not speak its name. The integrity of Antonio's body, though threatened, remains intact and disciplinary reformations are not necessary. By the end of the play, the audience knows that there will be no more trouble from this wether. Antonio has been disciplined and silenced, as has Shylock.

When Shylock is compelled to remain radically Other, perhaps almost, but, as Seymour Kleinburg comments, "not quite human" (121), one of the things that reinforces this position within the "fictional" world of *The Merchant of Venice* is the over three dozen references that compare him to dogs. He is a "cut-throat dog" (1.3.111) to Antonio; he is "the dog Jew" (2.8.14), "the most impenetrable cur / That ever kept with men" (3.3.18–19), according to Solanio; he is a "damned, inexecrable dog" (4.1.128) with a "currish spirit" (133) whose "desires / Are wolvish, bloody, starved, and ravenous" (137–38), in the eyes of Gratiano. He is "an alien"[20] and a beast. He accepts both positions. There is no question that he is "an alien," and Shylock himself confirms his status of dog: "Since I am a dog," he warns, "beware my fangs" (3.3.7). He would sink these fangs into the flesh of that "wether" (4.1.114) Antonio if he could. Shylock—a monster (partially humanized), a hybrid, a dangerous alien—must, like Caliban, be pruned and reformed, owned, brought into the fold, and with the terror of corporeal violence so imminent, he converts to Christianity, as much a corporeal conversion as any, a materially manifest correction of imagined deformity.

Portia herself, though she seems firmly housed in the center of power, is no less a hypervisibly (though not unproblematically) available object and commodity that meets the textual requirements for a compulsively expressive heterosexuality. She is linked with an irresolvably binaristic mineral world, at one and the same time being the lead and the gold, in the position of a passive object waiting to be chosen/mined by the right man.[21] The implicit equation of the woman with the mineral world is significant: minerals and women serve as objects of exchange between men. What I am interested in here in early modern discourses, particularly in the Elizabethan period, is less the gender-neutral metaphor linking the body politic with the body human[22] than the gender-specific equations of the female body with discursive productions of the environment. The conflation of the female body with the geographical body is very evident in the many maps of the period that cartographically inscribe women.[23]

If the New World Other is a part of the world that is objectified, commodified, and inventoried, so too is the Old World Other. There is nothing innocent about the gendering of early modern maps. As lands are divided up, so are women's bodies. Objectification, Carol Adams observes,

> permits the oppressor to view another being as an object. The oppressor then violates this being by object-like treatment. e.g., rape of women that denies women freedom by saying no, or the butchering of animals that converts animals from living breathing beings into dead objects. This process allows fragmentation, or brutal dismemberment, and finally consumption. (47)

Patricia Parker maintains that the discursive fragmentation of the female body enables the expression of a "lexicon of merchandising" (130), that "the 'inventory' of parts becomes a way of taking possession," and that the woman in question becomes a "gendered sign of territory to be conquered and occupied" (*Literary Fat Ladies* 131).[24] In *Twelfth Night*, Olivia itemizes herself and says that she shall be

> inventoried, and every particle and utensil
> labell'd to

my will: as, *item*, two lips, indifferent red; *item*,
    two
gray eyes, with lids to them; *item*, one neck, one chin,
    and so forth. (1.5.246–49)

Although her tone is ironic and dry, she does play out a familiar
theme—namely, the discursive division of the female body as a
passively available (and butcherable) object. In conceptual terms,
we find equations between figures of difference and the land; in
material terms, women are raped and butchered like the land;[25] in
terms of Elizabethan drama, women are, as Woodbridge scrupu-
lously points out, at times portrayed "as food, or as animals, or
as marketable commodities" (*Women* 262).[26] Of course, food and
animals are commodities, just as much as gold and rare metals
are. The production of the hypervisibly available object meets felt
needs.

In a Europe obsessed with marking boundaries, the denial of corporeal limits for the bodies of women and other figures of difference
strategically extends such bodies into a different set of spatial configurations than those of human communities.[27] It is a space that can
be known, mapped, divided, controlled, and mined. It is the natural
environment, which, as Robert Elliot notes in *Environmental Ethics*,
"possesses neither rights nor inherent value" (15). The environment
is morally considerable only in terms of its aesthetic or commercial
value. Moreover, we are probably all familiar with what Gillian Rose
identifies as a "complex discursive transcoding between Woman and
Nature" (88), if not in theoretical terms, then in the untheorized
terms of our daily experience. We have all heard terms like "Mother
Earth," "womb of the earth,"[28] and "virgin lands." Although there
are some gender-neutral metaphors that compare the earth to the
body ("bowels of the earth" is one), the majority of the metaphors for
nature are feminine.[29] The female body in early modern discourses,
though, is blurred in the moment of its articulation. It is a collection
of dismembered parts putatively *re-membered* but ultimately lacking firm borders. And a feminized body such as Antonio's, a body
unmanned by monstrosity such as Caliban's, a body cast out of the
rights and privileges of manhood such as Shylock's, and a body cast
away from humanity through madness: these each become subject
to the similar kinds of disciplinary action. To use modern parlance,
misogyny, homophobia, racism, and ecophobia converge. Staging
exotica within a spatial ordering outside the human sphere and into
the discursively deadened world of resources (mineral, botanical, and

animal) relies on and reproduces dynamic correlations between social and environmental domination and commodification. Ecophobia touches and influences many things. Conceptualizations of the environment are very much a part of early modern and modern processes that produce difference, and ecocriticism is just beginning to map this brave new theoretical world.

8

THE ECOCRITICAL UNCONSCIOUS: EARLY MODERN SLEEP AS "GO-BETWEEN"

Variously conceptualized as bestial passivity, dangerous inattention, sweet restoration, concord, preservation of life, imitation of death, the product of an insomniac mind, and, among other things, a wildly inconsistent motif, sleep in Shakespeare is bafflingly mercurial in early modern thinking, something that goes between different conceptual spaces, something that (for the maintenance of cultural stability) mediates between agonistic time categories (night/morning), themselves ideologically fascinating. Sleep, a go-between, a mediator of the very categories of "the human" and all that lies beyond, is one of the great unattended driving forces in the mapping of early modern culture—and it is an ecocritical issue acutely present in Shakespeare.

Although sleep as either a thematic or a theoretical issue certainly seems an unlikely candidate for ecocritical readings of Shakespeare, there are three general categories that need to be addressed: (1) Sleep intimates bestiality and thereby generates considerable literary representations of antipathy; (2) Diurnal sleep is seen and represented in Shakespeare as disturbing humanity's place in nature's order; and (3) Night and darkness (the proper covers of sleep) are consistently imagined as the flipside of everything good in nature, indeed of much that constitutes an abhorred nature.

AGAINST SLEEP

The prevailing morbid fear of impotence before natural forces that sleep represents in Shakespeare's England stems from the common belief that sleep was a cross-over into the animal world.

A direct and intimate link to nature, thus, sleep represents in many ways unpredictability and the kind of loss of agency and control that so often generates ecophobia. The most horrifying intrusion of nature into human affairs—affairs that are mistakenly perceived to be separate from nature—is, of course, death. Although there are many things about nature and natural processes that we can control, obviously death is not one of them. Early modern understandings of sleep as a kind of death are very telling witnesses to the belief that sleep is a go-between, a mediator of the very categories of "the human" and all that lies beyond.

Michael Sparke's 1638 comment that "sleepe [is] a kind of middle thing betweene death and life" (24) echoes in Lady Macbeth's comment that "the sleeping and the dead / Are but as pictures" (2.2.56–57). That sleep was a close sibling to Death was, according to Jean Robertson, "a commonplace from the time of Homer" (140), and it remains so in the early modern period. Sparke continues, but with growing antipathy toward sleep, arguing that "sleepe is proper to the body, not the soule, (for even then are we to be awake in soule, when wee sleepe in body) so dieth man in respect of his body, not his soule" (25). The idea here is that the mind is above the sordidness and compunctions of nature, the body dying while the mind remains alive (if not alert), or so the rhetoricians would have it.

The antipathy toward sleep recurs famously in *Hamlet*, when the eponymous hero speaks with contempt about sleep, asking "What is a man / If his chief good and market of his time / Be but to sleep and feed? A beast, no more" (4.4.33–35). The connection here between sleeping and feeding with animality—and, therefore, with the danger of boundary-blurring—could not be more clear.

Certainly in *Hamlet*, as George Walton Williams notes, "One might go so far as to say that the image for the sleeper is the beast" ("Sleep in *Hamlet*" 18). The interstices opened by the obsession with marking the human/animal, culture/nature boundaries in the Renaissance are filled with the sometimes lurid, often animalized, and always fickle go-between that sleep came to embody. David Bevington maintains that "sleep becomes a more ambiguous state in Renaissance drama than in its earlier manifestations; it grows more difficult to 'read' as a theatrical signifier" (53), and, indeed, the various Renaissance definitions of sleep are multiple, ranging from claims that it is nothing but an imitation of death to it being "one of the chiefe poynts of well ordering and gouerning ones self: concerning which there are certaine generall rules to be observed of them which are desirous to keep back and hinder the hastie access of old age" (du Laurens, cited by Dannenfeldt 422), "death's counterfeit" (*Hamlet* 2.3.76), or "chief nourisher in life's feast" (2.2.37). So, while it is neither always nor everywhere an indicator of animality, sleep, as a scapegoat or as a preserver, is often outside the troubled (and heavily smeared) boundary that defines the "human."

At one and the same time "swinish" (*Macbeth* 1.7.67) and a "Balm of hurt minds, great natures's second course" (2.2.36), sleep is a natural thing. The stark contrast between the representations of sleep is, in some ways, a reflection of the acute ambivalence toward the natural world, a world beckoned both to validate and to invalidate epistemologies, as the occasion demanded; but, as, Jean-Marie Maguin has so meticulously (if a bit arithmetically) argued, it is beyond question that

> Shakespeare's interest in night is constant and manifold. He uses it for comedy and tragedy alike. The volume of nocturnal action per type of play is telling: 2 percent of nocturnal action in the romances, 12 percent in the history plays, 16 percent in the comedies. The figure climbs to 25 percent for the tragedies... tragedy appears to have a natural affinity for it. This close accord between night and tragedy is no doubt an anthropological feature. (248)

But exactly what is it that accounts for such a high association of nocturnal action and sleep with tragedy?

Joseph Meeker's *Comedy of Survival* goes a long way to helping us answer this question. It predates but leans heavily toward ecocriticism and has recently received substantial and positively glowing reevaluations from the ecocritical community, largely (and rightly) on the grounds that it was far, far ahead of its time. Written in 1972, it argues that "literary tragedy and environmental exploitation in Western culture share many of the same philosophical presuppositions" (24). Meeker maintains that tragedy is a genre that strongly affirms the status of humanity over the natural world, a world which the genre figures as hostile. Trademarks of tragedy are impotence before nature, a persistent inability to conquer, subdue, and maintain control over nature. We see this, for instance, repeatedly in one of the greatest of all tragedies, *King Lear*, where we are lead dramatically into a profound fear of wild natures, human, and nonhuman. The height of individualism in Shakespeare's tragedies marks the height of anthropocentric thinking and desires for control of nature. Sleep is nature's incursion—and a very intimate one—into the lives of humans, a bestial incursion with tragic potentials. In many ways, too, the bestial implications of sleep are indeed a prerequisite of the suffering tragic subject.

Dinesh Wadiwel has recently made the connection eloquently clear: the "greatest acts of violence against humans appear to be accompanied by a *dehumanisation* that is of commensurate intensity" (1, emphasis added). Inappropriate sleep bestializes. Yet, we perhaps do well to contest the notion of "dehumanization"; if the subject of tragedy is generally stripped to the core (as Lear certainly is), tragedy supports Georgio Agamben's notion that "the animal" is not outside of but at the core of humanity,[1] that "the animal" is the skeleton that must be clothed in human flesh, as it were. Either way, whether "the animal" is imagined outside of or at the core of humanity, it is an undesirable site to which "the human" is loathe to go, one, though, to which the tragic subject is conveyed, and not infrequently through sleep.

It would be a gross misrepresentation of the facts, however, to suggest that only tragedy dramatizes a philosophy against sleep. Even a more comic play such as *The Tempest* echoes the dangers of sleep (or, more precisely, of the things that can happen during sleep): Ariel warns

> If of life, you keep a care,
> Shake off slumber, and beware.
> Awake, awake! (2.1.303–5)

Richard Levin has recently put the case very simply: sleeping in general was dangerous because "it left the napper vulnerable" (13).

Sleep's removal of agency makes it a useful stage device: Jennifer Lewin notes that "When rhetorical persuasion or physical violence cannot succeed, putting someone to sleep is always a viable, albeit temporary, solution" (186). Indeed, there is no shortage of examples of "the sleeping victim" on Shakespeare's stage, as David Roberts points out in his meticulous list (235). Sleep was a fascination to Shakespeare, no doubt because the loss of agency in the sleeper put all of the power in the wakeful. On stage, this puts a lot of power and responsibility into the audience. Writing sleep means writing a kind of voyeurism of somnolence. "Watching sleep," according to Roberts, "was a matter of almost obsessive concern to Shakespeare. No dramatist represents the act of sleep more frequently or graphically than he does" (ibid). And the power relationship between sleeper and wakeful is rarely innocent in Shakespeare. Rarely are the representations laudatory of sleep; the contrary is more often the rule. At best, suspicious things happen when people sleep in Shakespeare; at worst, down-right malevolent things occur.

Antipathy toward sleep notwithstanding, though, sleep remains nature's vehicle, over which people remain largely impotent. Sir Spencer St. John's comments of 1908 remain pertinent: "Nature effects different purposes through the agency of sleep [in Shakespeare]: sometimes, as in the case of Banquo, stimulating to evil; more frequently, as in the case of Macbeth, inflicting chastisement for evil committed" (196). The key word here, though subtly understated, is "nature."

AGAINST DIURNAL SLEEP

No less in Shakespeare's day than in our own, sleep is commonly held to be the natural property of night, not day, that it only happens diurnally by mischance, special circumstance, or more sinister and unnatural reasons. Henry the Fourth laments having offended a personified and gendered sleep, calling her "Nature's soft nurse" (*2H4*, 3.1.6–7), and wishing for the normalcy and naturalness of sleep to fall upon him at night, the time when so many of his subjects are asleep. It is a very common theme in Shakespeare that sleeping during the day is bad and goes very much against nature.

George Walton Williams maintains somewhat more gingerly that "there is some indication that sleeping by day has pejorative associations in the plays" (195). Williams argues extensively, noting that Shylock faults Launcelot Gobbo because he "sleeps by day" (2.5.45); that the Old Hamlet goes to sleep in the afternoon (1.5.59–60), with bad consequences; that Alonzo in *The Tempest* nearly loses his life by sleeping in the afternoon; that Falstaff, "wrong in many things...is also wrong in taking sleep in the day...is notable for his use of the night as the time of wakefulness and his keenest activity" (Williams "Shakespeare's Metaphors" 196), and is, in short, guilty of "sleeping upon benches after noon" (*1H4* 1.2.3–4); that there are numerous examples of the corollary view that it is wrong to work at night; and that Hotspur also makes inappropriate use of the night by riding (*2H4* 5.3.129; *1H4* 3.1.140) (196). And Shakespeare would have been well familiar with the Hippocratic ideal that people should " 'follow the natural custom of being awake during the day and sleep during the night,' and any change in this pattern was a bad sign" (Dannenfeldt 417).

Whether or not Shakespeare had read Avicenna, his works were certainly in circulation and having influence in early modern England.

One of the gems from Avicenna for early modern sleep doctors was
the argument that

> it was not good to sleep during the day, for this brought on illnesses
> associated with humidity, healthy coloring was lessened, the spleen
> became heavy, the sinews lost their tone, vim and appetite were lost, and
> fevers often appeared. If a person was accustomed to sleeping during the
> day, this practice should gradually be eliminated. (Dannenfeldt 419)

The idea—and it was very commonly held—was that sleeping during
the day meant going against nature, which meant nothing good. Just
how common was this view?

There is a surprising abundance of writing on the subject. The
prolific writer and physician Andrew Boorde maintains in 1562 that
"men, of what age or complexyon soever they be . . . , shuld take theyr
natural rest and slepe in the nyght & . . . exchew merydyall slepe"
(*Regyment* 246). Similarly, in 1599, the influential physician Andrea
Du Laurens urges, "Let euery man watch well ouer himselfe, that
he use no sleepe at noone. . . . It is good (saith Hippocrates) to sleepe
onely in the night, and to keepe waking in the daytime. Sleeping at
nooneday is very dangerous, and maketh all the body heauie and
blowneup" (157; 189–90). There is no question that "there was
general agreement" as Dannenfeldt claims, "that night-time was
the best time to sleep" (424) and that, to borrow a phrase from
Keith Thomas, "It was bestial to work at night" (39) in early modern
England.

The invocation of nature as the final arbiter for nocturnal sleep
in early modern thinking finds expression in Tobias Venner's 1637
argument that

> we must follow the course of Nature, that is, to wake in the day, and
> sleep in the night: *Dies enim vigilae, nox somno dicata est.* For the Sun
> by his radient beames illuminating our Hemisphere, openeth the pores
> of the body, and dilateth the humors and spirits from the Center to the
> circumferent parts, which to waking and necessary actions doth excite
> and naturally provoke. But on the contrary, when the Sunne departeth
> from our Hemisphere, all things are coarctated, and the spirits return
> into the bowels and inmost parts of the body, which naturally invite to
> sleep. Wherefore, if we pervert the order of Nature, as to sleep in the
> day, and wake in the night, we violently resist the motion of Nature,
> for sleep draweth the naturall heat inward, and the heat of day draweth
> it outward, so that there is made as it were, a fight and combat with
> Nature to the ruine of the body. (270–71)

Indeed, to cite Richard Levin again, sleeping during the day "was regarded not only as abnormal and undignified but also as somewhat immoral (a violation of the emerging Protestant 'work ethic')" (13). If sleep is associated with evil and evil nature, no less so are night and darkness.

AGAINST NIGHT

Much of the villainy and otherness of early modern literature depend for their full effect on the presence of an essentialized understanding of color, a chromotypic essentialism that defines residential evil in anything from clouds, yew trees, asses, dogs, and crows, to people, seasons, and hours. In John Webster's *The White Devil*, evil resides in nocturnality, and because women are repeatedly associated with night, the essentialized understandings of night and of women form a kind of joint subjectification; moreover, since women are also associated with animals in the most unflattering of ways, it seems remiss to neglect ecocritical commentary.

In *The White Devil*, whatever "blurring [of] the play's black/white, good/evil polarities" (Webster, ns.d.2–3, 52–53) occurs, unpleasant associations attached to blackness fall thick and heavy throughout: we get "black lust" (3.1.7); "black deed" (5.3.251; 5.5.12; and 5.6.300); "black concatenation" (3.2.29); "black Fury" (5.6.227); "black storm" (5.6.248); and "black charnel" (5.6.270). Francisco says in defense of Vittoria "I do not think she hath a soul so black / To act a deed so bloody" (3.2.183–84). Monticelso keeps a "black book" (4.1.33) in which he keeps "the names of all notorious offenders / Lurking about the city" (4.1.31–32); he calls Lodovico "a foul black cloud" (4.3.99) and talks about "the black and melancholic yew tree" (4.3.120); Flamineo, in disguise as "Mulinassar, a Moor," says he loves Zanche, "that Moor, that witch" (5.1.153) and several times compares her to a dog, complaining that "women are like / cursed dogs; civility keeps them tied all daytime, but / they are let loose at midnight" (1.2.196–98). Of course, this further entrenches the link between women and nonhuman animals and the danger that they represent at night if allowed liberty. At other points, Marcello compares Zanche to "crows" (5.1.196); Zanche says that she disliked her blackness until she met Mulinassar (5.1.213); and "eternal darkness," Vittoria explains, "was made for devils" (5.6.63–64)—in short, images of darkness invoke the natural environment in the process of conveying meaning, and in so doing produce the natural environment as both an active participant in and passive victim of racism. In other words, invoking the natural environment as a discursive resource for vilifying blackness means including the natural environment in the project of racism; it also means producing the environment as an object analogous to the vilified black Other.

The writing of nocturnal alterity has implications for discussions about environmental ethics, and there are heaps of examples of this kind of writing. In *The White Devil* alone, there are several noteworthy instances. We learn, for instance, that "your melancholic hare / Feed[s] after midnight" (3.3.82–83), and we see a linking of nighttime life with madness, lechery, and bestialism. When Flamineo asks Lodovico if he had "to live [like] a lousy creature . . . Like one / that had forever forfeited the daylight" (3.3.116–17), we understand that night life is undesirable. We might note that the othering of nocturnicity is still an issue, since the business world is geared to people who live diurnally. Nocturnal people have fewer entertainment opportunities, their shopping choices are often limited to expensive convenience stores, their jobs tend to be undesirable, and so on.

One of the world-inverting effects of 9/11 was that it happened in broad daylight. We do not expect evil under the garish eye of day. If, as Vittoria explains in *The White Devil,* "eternal darkness was made for devils" (5.6.63–64) in the early modern period, nighttime in the twenty-first century is no less the time for evil. Contemporary cultural representations of villainy seem in many ways to be associated with night: the Batman movies occur almost entirely at night (and we might note that the eponymous hero is a bestial hybrid— half human, half bat—ideally suited to fight the bestial crimes of the seemingly always dark Gothic City), Christian mythology never mentions a nighttime in heaven, ghosts and vampires do not come up with the sun, and so on. The villains of cultural imaginations early modern and present in the West often do their work at night and sleep during the day: and as we have seen, sleep during the day is represented in terms of unnaturalness as a thing that cannot lead to any good.

Night and the surrealism sleep engenders are also associated with witches throughout the early modern period generally and Shakespeare specifically. *Macbeth*'s witches, to take just one example, are creatures of the night that seem immune to the need for sleep. Indeed, it is through the witches that Macbeth comes to "murder sleep" (2.2.35), comes to kill "the death of each day's life" (1.37), producing a metaphor that turns in on itself in an absurdly unnatural way.

The deformities here are as much physical as conceptual, disruptions on many levels of natural order and human thinking, with all

THE ECOCRITICAL UNCONSCIOUS119

of the bizarre implications that killing the death of day (sleep) might imply.

IMPLICATIONS

Though sleep seems an unlikely object for ecocritical readings of Shakespeare, sleep (both as a theoretical and thematic issue) connects with ecocriticism in several important ways. Authors have written against sleep because of the bestiality it evokes. Writers have railed against diurnal sleep because it is thought to disrupt natural order. And the associations of night, darkness, and evil with sleep have filled the pages of early modern literature, with strong implications for how we understand early modern environmental ethics.

For ecocritical theorists, discussion of sleep in the early modern period is a very far cry from an activist ecocriticism, very far from the kinds of interventions against the very real and violent depredations that ecocritical readings seek to make. For all intents and purposes, discussions such as the ones offered in this chapter perhaps amount to little more than thematic ecocriticism and perhaps fall far short of the activist position that characterize the embryonic stages of ecocritical endeavor and, indeed, much of my own work. Thematic ecocriticism, though, must be defensible insofar as it offers foundational work on conceptual connections. Regarding sleep, both theoretical and thematic discussions about sleep have been sparse to date; have largely faltered because they have left unexplored the conceptual interfaces among such topics as race, night, crime, and safety; and have looked instead at image patterns, authorial dexterity, and matters of artistic cohesiveness. I have sought here to explore this largely unexplored in-between space of sleep more with an eye to making ideological connections than to foregrounding matters of imagery and thematicism (both of which, however, remain important both in the above discussions and in work yet to be done with sleep). I have sought to make the kinds of connections that ecocriticism should be making.

There is a growing and measurable dissatisfaction with ecocriticism, the connections it does not make, the methodologies it does not develop, and the promises it does not keep. Recently, Timothy Morton has argued that a lot of what claims to be ecocriticism is really a movement back to a different, earlier kind of criticism, that "just as the Reagan and Bush administrations attempted a re-run of the 1950s, as if the 1960s had never happened, so ecocriticism promises to return to an academy of the past" (20). In fact, many critics have suggested that ecocriticism has become self-satisfied; that ecocriticism may "end up being what it threatens currently to be, a new niche of professionalism, an easy place to publish and establish a name" (O'Dair "Is it...?" 85); that ecocriticism is becoming "one of those trends that temporarily guarantee an audience, publications, tenure, promotions, and so on" (Estok "Letter," 1096); that environmentalist issues are "an academic resource capable of providing novel exemplars for tired arguments or revitalizing flagging careers" (Smith *An Ethics of Place*, 3); that mainstream ecocriticism is little more than "a form of postmodern retro" (Morton 20); that ASLE has become a clique and that ecocritics have fallen asleep at the wheel; that it has "from its inception...adopted a belligerent attitude towards critical theory" (Parham 25); and that

> some ecocritics have made a point of expressing their distaste for theory in language that suggests an impatience not only with theory but also with any intellectual activity trafficking in abstractions, as if ecocriticism needed no definitions, as if it could begin and end by praising the objects of its attention—as if ecocriticism were to be organized and run as a sort of fan club. (Phillips 138)

Indeed, Michael Cohen writes similarly, arguing that "in its enthusiasm to disseminate ideas, a certain version of narrative ecocriticism might better be described as praise than criticism" (21). And there are very serious dangers in the "praise song" school of ecocriticism: "The complacency of the praise songs and the denial of real contesting positions," Cohen argues, "will mean slow stagnation" (23).

Doing ecocriticism means being attentive, alert, and alive. Doing ecocriticism with Shakespeare in addition means, as Frederick Waage observes, being "held to higher standards than 'other kinds of theorists'" (140). Doing ecocriticism with Shakespeare means keeping alive the radical potential of ecocriticism, keeping alive the promises ecocriticism makes to connect. Ecocriticism is potentially radical, if it does what it promises. The breadth and potential reach of its analytical scope, with its various commitments (social and environmental) make ecocriticism very different from other theories. Doing ecocriticism means facing those commitments.

Doing ecocriticism with Shakespeare means doing it with an eye to what was going on at the time, and this means seeing (not inventing) radical connections and explaining them. Part of this for ecocriticism has meant also using tools that are new. Ecocriticism is one such family of tools, and like screwdrivers that have different species, ecocriticism has different branches. Theorizing ecophobia, which has been at least a part of my project here, is one of the branches of ecocriticism, one that helps us to articulate a methodology that is both useful and necessary.

For Shakespeareans, the very mention of ecocriticism has brought strong responses—not only of interest. Perhaps the higher standards of which Waage speaks are a valid requirement, a requirement that spells the difference between purely thematic criticism that often comes very close to the work of ecocriticism on the one hand and work that actually does what ecocriticism seeks to do on the other. Take, for example, Jennifer Lewin's comments that "sleep can represent either the vulnerable lure of oblivion or the eerier attractions of seizing control over one's circumstances" (190). While there is nothing untrue here, the comments, though they come tantalizingly close to making radical connections, in fact, fail to do so. An ecocritical reading, far from rejecting Lewin's insights, however will build on them. In such a spirit, we might argue further then that both scenarios Lewin notes are removals from the control that characterizes the rationalist pursuit of power and the humanist pursuit of realizing the self-governing individualist ego, pursuits increasingly characteristic of early modern values, values contingent on a rejection of nature's autonomy and agency in favor of control. Sleep, simply put, threatens that control. The "higher standard," then, of which Waage speaks, is perhaps to be expected, since ecocriticism seeks a much more sophisticated and developed set of responses than the "other kinds of theorists" he mentions.

This chapter has sought those responses, to prove that sleep—and the ecophobia that supports and is supported by its representations in the early modern period—is clearly a complex issue, one acutely implicated and imbricated with discourses about environment.

Coda: Ecocriticism on
the Lip of a Lion

Printed on stone high above the books and heads of people in the Social Science Reading Room at the Library of Congress in Washington, DC, one can read that "the earth belongs always to the living generation. They may manage it then and what proceeds from it as they please during their usufruct." Such is, the writing in stone continues, one of the cornerstones of human freedom. Yet, we all know the environmental horrors to which such a principle has lead. We have all heard the apocalyptic prophets arguing for our doom and the scientists arguing for our salvation. We have all heard the debates about our changing weather, our increasingly tainted food, air, and water, and our diminishing resources. We know the arguments, but nothing seems to be changing in the way we relate with the natural world. Things only seem to be getting worse. Why this is so has to do with how we understand our most cherished ideals.

Critiquing Western environmental ethics in some ways means critiquing the imagined bases of hard-won freedom and democracy. It means critiquing the parameters of freedom and the parameters of democracy—in effect, critiquing the imagined essences of each and at times perhaps even engaging in a balancing act between civil liberties on the one hand and environmental respect on the other. In some ways, it means abandoning certain concepts of personal rights, and in others, it means extending them to the nonhuman world. It means envisioning the "democracy extended to things" (12; 142) Bruno Latour speaks of. It means taking things personally and making personal changes. And it means making connections.

Studying how someone such as Shakespeare connects with the present environmental crises that we daily breathe and smell and eat and taste, the difficulties and tragedies we live through and cause, does several things. It forces us to imagine the literature and the theory through new perspectives, to examine complementary systems of thought, and to develop a vocabulary for concepts that have no names. It also allows us to define more fully the goals, methodologies, and terms of ecocriticism.

What counts as ecocriticism is any theory that is committed to effecting change by analyzing the function—thematic, artistic, social, historical, ideological, theoretical, or otherwise—of the natural environment, or aspects of it, represented in documents (literary or other) that contribute to the practices we maintain in the present, in the material world. To reiterate O'Dair: "Ecocriticism of Shakespeare is presentist" ("Is It Shakespearean Ecocriticism If It Isn't Presentist?" 85). Ecocriticism embraces possibilities. It is committed to diversity and to innovation, to imagining possibilities, and to hope. Applied to Shakespeare, it extends huge amounts of foundational work that has already been done into exciting new areas. It can help us to understand where we have come from, where we are, and where we might be going.

It seems obvious and barely worth arguing at this stage that there is a fundamental difference between ecocritical readings of Shakespeare, on the one hand, and, on the other, the volumes of very useful scholarly work that have been produced over the centuries about representations of nature in Shakespeare, a difference, in other words, that Sharon O'Dair has characterized as "old school 'nature studies' and new school 'ecocriticism'" ("State" 476). This much seems obvious. It also seems obvious that what has become a defining focus of ecocriticism is central to each and every chapter in this book—specifically, an emphasis on the real, the material world we daily breathe and smell and feel when we walk outside, the world that rains on us, starves or feeds us, drowns or burns us, the world we reconstruct through discourse, a world that nevertheless exists before our discursive constructions of it, and will no doubt exist in some form or another long after we are gone (and it is just a matter of time before we *are* gone). What is less obvious is why, if we are really concerned about the environment, we should bother with Shakespeare—seems at best a bit of a stretch to connect this old dead guy with current *environmental* crises, of all things. Better perhaps to join *Greenpeace* if we really want to be activist. Ecocritics are a puny minority in world affairs: an average movie attracts more attention than all of the ecocritics combined ever will. A cynical view might have it that those doing this puny new -ism with an old and established giant such as Shakespeare are like parasites, like the "flea that dare eat his breakfast on the lip of a lion" (*Henry V*, III, vii, 145–46).

My intention throughout this book has been to suggest possibilities and to open doors, rather than necessarily provide definitive answers.

A viable ecocriticism has little future without, to some extent, closing some borders (whose unrestricted openness have become ambivalent, a liability) and, to some degree, restricting entry; similarly, the kind of future a feminist criticism would have by nominating Hugh Hefner a feminist would be at best questionable. Producing a viable ecocriticism means adopting a tacit intellectual understanding that if it is sexist, then it cannot qualify as ecocriticism, since sexism goes against the spirit, goals, and vision of ecocriticism. This need not mean that ecocriticism need always be actively feminist—it would be nice, but that might be asking a bit much. Similarly, if an environmentally oriented critique is demonstrably racist or homophobic, then, again, it cannot qualify as ecocriticism, for the same reasons—and, again, the criticism need not always be actively and demonstrably seeking an antiracist or antihomophobic project, but it does always need to be *not* promoting racism, or bigotry based on cultural, ethnic, religious, sexual, or geographic grounds. The more that ecocriticism does theorize itself in confluence with other activist theories, the better off it will be. Using Shakespeare and his representations of a certain breed of ethics toward the natural world (which I have been theorizing under the label "ecophobia") can produce the basis for such an ecocriticism that theorizes in confluence, and it has been the goal of this book to do so. Ecophobia is certainly not the only ethical paradigm Shakespeare represents, and I do not on any level want to imply that it is. Ecocriticism and the paradigm of ecophobia, similarly, do not and cannot give us all of the answers; but each can help us enormously in moving toward them.

NOTES

1 DOING ECOCRITICISM WITH SHAKESPEARE: AN INTRODUCTION

1. Since 1998, when the first published instance where the words "Shakespeare" and "ecocriticism" appeared together (see Estok, "Environmental Implications," *Shakespeare Review* 33, p.135, n.39), the field of "Shakespeare and ecocriticism" has become flooded with scholarship. To date, there have been two books published that apply ecocriticism to Shakespeare: Robert Watson's *Back to Nature: The Green and the Real in the Late Renaissance* and Gabriel Egan's *Green Shakespeare: From Ecopolitics to Ecocriticism.* Articles (now too many to number) and special issues (notably, the "Shakespeare and ecocriticism" special cluster of *ISLE* in the summer of 2005) have been increasingly appearing. Palgrave Macmillan published a collection entitled *Early Modern Ecostudies: From the Florentine Codex to Shakespeare* (Ed. Ivo Kamps, Thomas Hallock, and Karen Raber) in December 2008, and there is another collection forthcoming through Ashgate entitled *Ecocritical Shakespeare* (Ed. Lynne Bruckner and Dan Brayton) in 2011, as well as a full monograph by Todd Borlik entitled *Ecocriticism and Early Modern English Literature* set to come out through Routledge as this book goes to press. There have been several conference panel sessions over the past several years: the 2008 Shakespeare Association of America (SAA) held a session entitled "Shakespeare and Ecological Crisis" and a seminar in 2006 on "Nature and Environment in Early Modern English Drama." The 2006 World Shakespeare Congress (WSC) held a panel session organized by Simon Estok entitled "Ecocriticism and the World of Shakespeare," while the 2011 WSC has one organized by Vin Nardizzi and Jennifer Munroe entitled "Plants and Gender in Early Modern Literature." The British Shakespeare Association held a "Shakespeare and Ecology" seminar organized by Gabriel Egan and Kevin de Ornellas at the 2005 Biennial Conference, and the 2001 Ohio Shakespeare Conference entitled "The Nature of Shakespeare" sponsored several ecocritical sessions. There was a session at the 2008 MLA entitled "Ecosystemic Shakespeares," and panel organized by Sharon O'Dair at the 2009 ASLE Conference ("Island Time: The Fate of Place in a Wired, Warming World") held in Victoria, BC, entitled "Early Modern Ecocriticism—Three Ways," and a

Symposium organized by Scott Newstok entitled "Green Shakespeare: Environmental Criticism and the Bard" at Rhodes College, Memphis, in March 2010.

2. See Stolz (http://achangeinthewind.typepad.com/achangeinthewind/2005/11/ecophobia_a_par.html). It is one of the aims of my book to offer and extensively define the term "ecophobia." For the time being, we might define ecophobia as a pathological aversion toward nature, an aggravated form of anthropocentrism expressed variously as fear of, hatred of, or hostility toward nature at least in part motivated by a sense of nature's imagined unpredictability. A more detailed definition of the parameters and implications of ecophobia begins on the following pages and continues throughout the book with readings that will add more flesh to what must here seem skeletal.

3. This is not to imply, however, that membership in one category means membership in all or that being African American means necessarily being antihomophobic or that all vegetarians are necessarily antiracist. Indeed, as Sharon O'Dair has observed, "people can suffer injustice themselves and impose it on others—other people or other species" (Personal correspondence, "Notes" November 10, 2008). My point is that conservatism tends to go across the board (as does radicalism), that we tend to find ideologies huddling together, that "rednecks" who drive pick-up trucks with offensively conservative opinions about women tend also to have offensively conservative opinions about sexuality, race, environment, and so on. While there are obviously exceptions—for instance, gay cowboys, classist African Americans, conservative lesbians, and homophobic vegetarians—much is to be gained from discussing confluences.

4. Portions of this paragraph have appeared in earlier forms in "An Introduction to Shakespeare and Ecocriticism," *ISLE* (12.2), 112–13; "Shakespeare and Ecocriticism," *AUMLA* (103): 17–19; "Ecocritical Theory and Pedagogy for Shakespeare: Teaching the Environment of *The Winter's Tale*," *Shakespeare Matters: History, Teaching, Performance* (2003): 196, n5; and "Conceptualizing the Other in Hostile Early Modern Geographies," *Journal of English Language and Literature* (45): 878–90. Many of the ideas here also found expression in "Theorizing in a Space of Ambivalent Openness," *ISLE* (16.2): 203–25.

5. Dates here are important. My PhD dissertation was accepted and dated in the Spring of 1996. I wrote the first draft of the final chapter of my dissertation in the early summer of 1995 and submitted it to Linda Woodbridge—my supervisor—on August 9, 1995. Several months later, by which time the dissertation had already gone to my committee for approval, David Sobel's "Beyond Ecophobia: Reclaiming the Heart in Nature Education" came out in *Orion* (though it was not until shortly before the ASLE 2009 meeting in Victoria—about a month after my own article on ecophobia came out that Sobel's article came to my attention. The fact that Sobel and I clearly coined the same term at roughly the same

time and clearly independently is perhaps more than simply coincidental, perhaps indicating a felt need for a viable ecocritical terminology, as early as 1995.

6. Much of my thinking about ecophobia grows out of foundational works of early ecofeminists, particularly as Susan Griffin's *Woman and Nature: The Roaring Insider Her* (1978), Greta Gaard's collection entitled *Ecofeminism: Animals, Women, and Nature* (1993), and Irene Diamond and Gloria Orenstein's *Reweaving the World: The Emergence of Ecofeminism* (1990).

7. A reader for an earlier version of this book has complained to me that rape is not always about misogyny. Fair enough, but I doubt that any sane person will argue against the idea that rape *often* implies misogyny.

8. Although the contempt and fear, which I am calling ecophobia (and expand on below), does not represent the *sole* trait that characterizes our relationship with the natural world, it is as yet a remarkably unattended one. Its opposite would, to some extent, be the biophilia Edward O. Wilson defines as "the innately emotional affiliation of human beings to other living organisms" (31). Certainly Scott Slovic is accurate to note that "ecocriticism is actually motivated by biophilia" (Personal correspondence, "Re: LIKELY SPAM" September 16, 2008). Admittedly, biophilia indeed seems to be the *motivation* but not the *object* of ecocritical inquiry. The object of such inquiry certainly must centrally include ecophobia and how it patterns our relationship with nature. We can clearly see that ecophobia is winning out over biophilia. The "rapid disappearance" (Wilson 40) of species of which Wilson speaks so eloquently and persuasively has a cause: it is ecophobia, surely, not biophilia.

9. As I have mentioned in *ISLE* 16.2, the topic has become an increasingly marketable one, with the Animal Planet/Discovery Channel's joint production of the CGI series *The Future Is Wild* (2003), Alan Weisman's 2007 book *The World without Us*, the History Channel's *Life after People* (January 2008), and the National Geographic Channel's *Aftermath: Population Zero* (March 2008), each, in their own way, move beyond mere ecological humility, tacitly presenting an implicitly ecophobic vision of a Nature that will finally conquer humanity, will reclaim all of the world, and will remain long after we are gone. Tom J. Hillard, in a productive response to my comments in *ISLE* 16.2, has added that

> Anyone who has paid any attention to popular culture in the past decade or so has no doubt witnessed a dramatic rise in the number of such stories. For instance, the 2004 blockbuster *The Day After Tomorrow* presented an astonishing, special effects-laden vision of the world during an apocalypse brought about by global warming and climate change, a vision of disaster terrifying in its implications (even if, at times, laughable on screen). This film is just one in a long line of natural disaster movies that

flooded theaters in the late 1990s, including the likes of *Twister* (1996), *Volcano* (1997), *Dante's Peak* (1997), *Armageddon* (1998), *Deep Impact* (1998), and many others. More recently is *Open Water* (2003), based on a true story of two scuba divers left behind by their charter boat and eventually eaten by sharks, as well as *Grizzly Man* (2005), a documentary about Timothy Treadwell, who, after thirteen summers of living with grizzly bears in Alaska, was eventually mauled and eaten by one of them in 2003. (187)

10. In work that predates ecocriticism, several capable scholars have, in fact, discussed precisely the topic "the domination of Nature" (see, e.g., Leiss, Evernden, and Roszak).

11. The most recent and perhaps most disturbing manifestation of this resistance to theory comes from within the ecocritical community itself. Scott Slovic's decision to publish (in *ISLE* 16.4) an incendiary and divisive rant against my "ecophobia" article has put my call for theorization at the center of a growing debate about the place of theory in ecocriticism. In turn, responses to *ISLE* 16.4 itself have been so intense that by December 2009, Scott Slovic had felt compelled to consider "issuing a call for submissions to a special forum on the broader topic of 'Ecocriticism and Theory' that would appear in one of the 2010 issues of ISLE" (Slovic "Further Reflections").

12. See, for instance, John Evelyn's *Fumifugium: or, The inconveniencie of the aer and smoak of London dissipated*, which describes the air of London as a "thick mist accompanied with a fuliginous and filthy vapor" (5). See also Ken Hiltner, who brilliantly discusses the matter of "the manner by which industrial air pollution emerges as an issue of discussion in seventeenth-century England" (429), how "to broach a problem that no one wishes to acknowledge because everyone is in fact its cause" (430), and how very relevant these representational matters are *today* to "our role in contributing to the crisis" (436).

13. By 1676, John Graunt was writing that "the Smoaks, Stinks, and close Air, are less healthful than that of the Country" (63); that "The country is more healthful than the City...[and] the Fumes, Steams, and Stenches...make the Air of London...not more healthful" (94). The reason is less owing to the increasing population than to the fact that the fuel being used was polluting the air. Graunt argued thus:

> I considered, whether a City, as it becomes more populous, doth not, for that very cause, become more unhealthful: and inclined to believe, that *London* now is more unhealthful than heretofore; partly for that it is more populous, but chiefly because I have heard, that sixty years ago few Sea Coals were burnt in London, which are now universally used...many people cannot at all endure the smoak of London, not only for its unpleasantness but for the suffocations which it causes. (Graunt 94–95)

14. More often than not in the early modern period, legal acts to preserve the environment lacked an ecological ethic and were made more in the interests of industry than ecology. (The word "ecology" didn't even exist until 1873, though the concept it describes very likely did.) Henry VIII's *Act for the Preservation of Woods* was really an act for the preservation of the raw materials of industry, for, as the *Act* said, "unless speedy remedy in that behalf be provided, there is a great and manifest likelihood of scarcity and lack, as well as of timber for building, making, repairing and maintaining of houses and ships, and also for fewel and firewood for the necessary relief of the whole commonality of this his said realm" (cited by James 125).

15. For a thorough analysis of issues attending the diminishing supply of timber and the implications of this shortage for Britain's status as a naval power, see Robert Greenhalgh Albion. While Albion provides more a naval history than a history of environmental ethics, the book remains an impressive piece of scholarship. See also Todd Borlik, who has recently done some useful work in this area.

16. While both John Evelyn (1670) and Andrew Yarranton (1677) argued how industry might be seen to preserve forests, it would be a mistake to characterize them as having been oblivious to the growing problem of supply. In *England's Improvement by Sea and Land*, Yarranton maintained that various industries work "in preserving Woods for their continuation and duration" (60) and that, were it not for the iron-works, "all these Copices would be stocked up, and turned into Tillage and Pasture, and so there would be neither Woods nor Timber in these places" (Pt. 1, 60–61). The iron-works, in short, may "increase Woods" (Pt. 1, 147), if done properly, part of which for Yarranton meant creating more enclosures; but, motivation notwithstanding, Yarranton was very clear that "it is now high time to think upon some apt and quick means for the preservation and increase of Timber in all Copices woods throughout England" (Pt. 2, 77). Similarly, while Evelyn felt that iron-works may have been "a means of preserving...woods" (cited by James 123), in *Silva: or, a Discourse of forest-trees*, he was, despite his motive, clearly deeply concerned with preservation.

17. For an extensive list, see Raber.

18. The preceding three paragraphs appeared in slightly different form in my article "Doing Ecocriticism with Shakespeare."

19. It is indeed timely to discuss weather in *King Lear*, and work is beginning to appear on the topic from explicitly ecocritical perspectives (see, for instance, Markley).

20. When I first wrote this sentence in an article that appeared in 1998, there was nothing that had been done with the topic of monsters from ecocritical perspectives. Georgia Brown has a provocative and useful article on the topic entitled "Defining Nature through Monstrosity in *Othello* and *Macbeth*." It is a great beginning, but there is an enormous amount of work yet to be done theorizing monstrosity and ecocriticism, some of which chapter 5 below attempts.

2   DRAMATIZING ENVIRONMENTAL FEAR: *KING LEAR*'S UNPREDICTABLE NATURAL SPACES AND DOMESTIC PLACES

1. Historian Brian Fagan discusses the effects of cooling in the waters where the English sought fish, arguing (among other things) that if there were better historical records, "the changing distribution of cod in the far north would be a remarkable barometer of rising and falling sea temperatures" (Fagan 71).

2. Edgar, however, holds a different view about what unaccommodated men such as he owe to the natural world. Having just seen his blinded father in the storm, he claims that "the wretch that thou has blown unto the worst/ Owes nothing to thy blasts" (4.1.8–9). Whether or not this "nothing owing to the natural world" represents the environmental ethic of the play as a whole is impossible to say without straining at the seams of textual plausibility, but it does accord with the waning of organicism of the period in which the play was written.

3. It is surprising how little attention this important topic has received. Gabriel Egan has a sentence on "the Little Ice Age" in *Green Shakespeare*, but it is only with Robert Markley—who also notes that questions about early modern weather have been "all but unnoticed by most modern commentators" (131)—that we find the first extended discussion of the topic.

4. Jonathan Dollimore seems to hold a slightly different opinion on this matter, suggesting that Lear in fact is fully experiencing the situation he is in. Dollimore argues that

   (The distracted use of the abstract—'You houseless poverty'— subtly suggests that Lear's disregard has been of a general rather than a local poverty). He has ignored it not through callous indifference but simply *because he has not experienced it*. (191)

   The clear suggestion is that he *is* now experiencing it, but I would argue otherwise, since the second person abstraction points away from Lear rather than toward him.

5. The sensory appeal of Lear's experiences is perhaps nowhere more vivid than in the images he uses when he rails against what he sees as a conspiring Nature:

   Blow, winds, and crack your cheeks! rage, blow!
   You cataracts and hurricanoes, spout
   Till you have drench'd our steeples, [drown'd] the cocks!
   You sulph'rous and thought-executing fires,
   Vaunt-couriers to oak-cleaving thunderbolts,
   Singe my white head! And thou, all-shaking thunder,
   Smite flat the thick rotundity o' th' world!
   Crack nature's moulds, an germains spill at once,
   That makes ingrateful man! (3.2.1–9)

6. The Gentlemen to whom Lear says he is king do not confirm his statement but merely accede that "You are a royal one, and we obey you"
   (4.6.201). To them, the sight he presents is "most pitiful in the meanest
   wretch, / [And is] Past speaking of in a king" (ll.204–5). The space and
   circumstances Lear inhabits deny him the identity he craves.
7. Ursula Heise mentions the "attachments to different types of spaces"
   that often function as "an integral part" of cultural and national identities (5), what Yi-Fu Tuan calls "topophilia"— "the affective bond
   between people and place" (4). Ecophobia and topophilia (like ecophobia and biophilia, discussed briefly in the previous chapter) are in some
   ways at opposite ends of an ethical continuum that characterizes a binaristic relationship toward the physical environment. Like early modern
   women, demonized or idolized, the natural world is a site of profound
   ideological versatility.
8. Woodbridge's detailed comments about Cordelia in an argument about
   vagrancy are worth quoting at length here:
       The link between domestic and national homelessness is built into
       Lear's thinking, a product of Tudor ideology identifying nation
       with home. Disinheriting his daughter, Lear turns her into a for-
       eigner. He has "stripped her from his benediction, turned her /
       To foreign casualties" (4.3.44–45). Conjuring the legendarily bar-
       baric Scythian, he orientalizes her: "The barbarous Scythian, / Or
       he that makes his generation messes / To gorge his appetite, shall
       to my bosom / Be as well neighbored, pitied, and relieved / As
       thou my sometime daughter" (1.1.116–19). Which means not at
       all: Lear's Britain hardly provides foreign aid to Scythians or can-
       nibals. The word "neighbored" suggests an idealized world where
       neighbors spontaneously help the poor, a world before wandering
       beggars were exempted from pity. "Relieved" could suggest private
       charity, but a primary meaning of "relief" was "assistance . . . given
       to the indigent from funds administered under the Poor Law or
       from parish doles" (OED). To Lear, Cordelia is ineligible even for
       poor relief; she is among the undeserving poor. "Stranger," which
       he twice calls her (1.1.115, 207), often meant "foreigner," and
       was also the term parish registers used for a person not of the par-
       ish, and hence ineligible for poor relief—vagrants were strangers.
       Kent emphasizes her shelterlessness: "The gods to their dear shel-
       ter take thee, maid" (1.1.185), and even after the King of France
       takes her up, Goneril emphasizes Cordelia's precarious position in
       a world of changing fortunes and reliance on alms-giving: "Your
       Lord . . . hath received you / At Fortune's alms" (1.1.281–82).
       Lear has turned his daughter into a vagrant. (285–86)
9. Kent is consistent in his beliefs, claiming in the fourth act that "it is the
   stars, / The stars above us, govern our conditions" (4.3.32–33).
10. Hard-heartedness, empty-heartedness (1.1.153), and dog-heartedness
    (4.3.45) seem interchangeable in this play, interesting because such a

metaphoric nexus conceives of the nonhuman as amoral and empty, a space empty of conscience and decency, a space of nothingness.

11. Yet, Edmund contradicts himself when he adopts his melancholic air and talks to his brother about the "unnaturalness between the child and/ the parent" (1.2.144–45) that he pretends to fear will come. If we assume that this fictional character is a unified subject with unified and consistent opinions, then it seems difficult to believe that he can possibly feel that such strife is natural (at least in the sense of it being something desirable), but there is no reason to assume that he is a consistent character with consistent feelings. He is, after all, a son who has substantial grievances against his parents, of a sort that makes everyday adolescent defiance to parental authority pale. Edmund's defiance, we may assume, is no less ambivalent (and is perhaps more so) than are more garden varieties of defiance to parental authority.

12. Monsters, a constant threat to the category of the human, crowd the early modern stage and are essential to delineating the reach of the natural, as well as the moral and ethical limits of the human. Chapter 5 below discusses in detail monsters and their relations with environmental matters.

13. The psychoanalytic tradition that reduces sexism to genital bias is misled. In the early modern period, the patriarchal idea that the vagina is nothing is part of a much larger issue: misogyny. It is not confined to the genitals. It is not merely vaginophobia; misogyny is contempt for the whole woman, the whole body and mind. With Edgar, we see that women's will is an "indistinguish'd space" (4.6.271), a space of virtual nothingness, but a space of boundless danger like the wilderness of the natural environment. The ease with which literature deploys bestializing metaphors and environmental metaphors against women speaks to a set of material practices that is at once ecophobic and misogynist.

14. As Elizabeth Mazzola and Corinne S. Abate note, recent commentary from Stephen Greenblatt suggests that Edgar's disgust is best understood as an allusion to a "terror inspired by the shadowy regions of the *vagina dentate*" (1–2), to use their words.

15. Wynne-Davies is building on Foucault's discussion of sanguinous power relations. Foucault's account of the early modern period offers a view of a culture steeped in blood, as it were, with blood constituting "one of the fundamental values" and blood relations constituting "an important element in the mechanisms of power" (*History of Sexuality, I* 147).

16. A sermon, circa 1623, claims, as Karen Newman summarizes, that a "woman is a series of prosthetic parts" (9), and the body onto whom she is "fastened" is the husband. The sermonist says that

> the *Woman* that beareth the *Name*, and standeth in the *roome* of a *Wife*, but doth not the *office* and *dutie* of a *Wife*, is but as an *eye of glasse*, or *a silver nose*, or *an ivorie tooth*, or *an iron hand*, or *a wooden leg*, that occupieth the *place* indeed, and beareth the

Name of *a limbe* or *a member*, but is not truly or properly any *part* of that *bodie* whereunto it is fastened. A failure to comply with "dutie" seems at least to imply some kind of punitive discursive dismemberment here. In an argument predating eco-criticism, Newman points out that the woman here has been broken down into commodities—silver, hardwood, and ivory (12)—and is subject to merchandizing.

3 *CORIOLANUS* AND ECOCRITICISM: A STUDY IN CONFLUENT THEORIZING

1. Kenneth Burke argued famously in 1966 that "[i]n the light of Freudian theories concerning the fecal nature of invective, the last two syllables of [Coriolanus's] name are so 'right,' people now often seek to dodge the issue by altering the traditional pronunciation (making the *a* broad instead of long)" (147). Maurice Hunt, meanwhile, maintains that "in terms of the corporate imagery of the body politic in *Coriolanus,* the protagonist, nominally associated with the anus, is expelled" like feces (221). Jonathan Goldberg, meanwhile, explores what he sees as anal erotics in the play, arguing in ways that some have found offensive that "the oral and anal exist in a substitutive relationship in the play" (263) and that "it is time to open up the question of anal erotics" (264).

2. Under the topic of queer ecocriticism, while there has been no manuscript yet published dealing explicitly with the topic, there have been several articles. The first was Greta Gaard's pioneering "Toward a Queer Ecofeminism." Since that 1997 article, Catriona Sandilands, who has become one of the few lone voices queering environmental politics, is all-too-correct in asserting "that environmentalists haven't had much to say about heterosexism and homophobia" ("From Unnatural Passions" 31). Robert Azzarello echoes this concern in his June 2008 CFP "Queer Ecocriticism and Theory" (http://call-for-papers.sas.upenn. edu/node/23972), where he asks "Why has queer theory been so disconnected from environmental studies?" The book *Queering the Non/ Human* (2008), edited by Noreen Giffney and Myra J. Hird, though it is not explicitly about ecocriticism, is solid contribution to the field, as is Wendy Lynne Lee and Laura M. Dow "Queering Ecological Feminism: Erotophobia, Commodification, Art, and Lesbian Identity" (2001).

3. It needs saying at this point that although I use the term "homosexual" freely here, I use it with the same reservation Alan Bray uses it: "To talk of an individual in this period as being or not being 'a homosexual' is an anachronism and ruinously misleading" (16).

4. Peter Stallybrass offers an enlightening discussion of "when and how…the specific signifier 'individual' was deployed in England in the seventeenth century" (593), noting, interestingly both that Shakespeare

does not actually use the term and (referencing Raymond Williams on the term) that other uses of it in the period are opposing:

Individual originally mean indivisible. That now sounds like a paradox. "Individual" stresses a distinction from others; "indivisible" a necessary connection. The development of the modern meaning from the original meaning is a record in language of an extraordinary social and political history. (Williams *Keywords*, 133 as cited by Stallybrass 594)

Though the argument of both Stallybrass and Williams is only tangentially related to the current discussion of *Coriolanus*, the difficult conceptual separations between individual and community under discussion in each are central to the struggles with which Coriolanus wrestles.

5. Between 1568 and 1588, Elizabeth issued seven proclamations specifically against what she thought might constitute or encourage sedition (see proclamation numbers 114, 133–34, 137, 225, 229, 259, and 273a in *Proclamations, 1588–1603*). The nine proclamations (and one commandment) against excess apparel Elizabeth issued (see proclamation numbers 47–49, 52–554½, 65–66, 99–96, 154–57, 168–73, 196–201, 252–57, 343–47, and Com. 212), though they do not mention fear of sedition as their motivation, were interested in maintaining social order and decorum that the antisedition proclamations sought to maintain. See also Youngs, in particular 156–74.

6. See also Adelman and Shanker (1949).

7. Excesses of blood, as Gail Kern Paster explains (79–84), are a symptom associated or imagined in the early modern period with women. The excessive bloodiness of Coriolanus, combined with his wounds (perhaps symbolically vaginal), feminize him.

8. Bristol argues that "balanced views of *Coriolanus* are typically more concerned with Shakespeare as the source of a complex and aesthetically satisfying elaboration of conflict than they are with description of a singular character" (219). Of the two sorts of description, Jagendorf seems to prefer the latter.

9. When faced with his mother, he complains that "like a dull actor now / I have forgot my part" (5.3.40–41). Earlier, also facing his mother, he clearly states his awareness that what he considers his essential manhood might be simply a part he is playing: he asks Volumnia "Would you have me / False to my nature? Rather say, I play / The man I am" (3.2.14–16). Earlier, Cominius has prepared us for the possibility that Coriolanus performs his gender by choice, and that "when he might have acted the woman in the scene [of battle with Tarquin] / He prov'd the best man i'th'field" (2.2.96–97).

10. Of course, the difference is that in *Titus*, the cannibalism is literal: Tamora "daintily hath fed / Eating the flesh that she herself hath bred" (*Titus* 5.3.61–62); in *Coriolanus*, the cannibalism of an unnatural dam eating her children is merely metaphorical. Another difference is that unlike the Rome of *Titus Andronicus*, the Rome of *Coriolanus* does not

lack a head. But Volumnia is the real head, Coriolanus being merely titu-
lar. Volumnia is the voice; Coriolanus, a boy of tears who only does what
he does "to please his mother" (1.1.38–39).

11. Virgilia, with her "O Jupiter, no blood!" (1.3.38) and her wish for her
son to be unhurt, meanwhile, seems a bit out of place in this murder
of crows. She is the conduct book foil for Volumnia, the surrogate
father Menenius, Valeria, and the rest of the bloodthirsty Roman
family.

12. We know from Cominius that at sixteen,

> with his Amazonian chin he drove
> The bristled lips before him. He bestrid
> An o'erpress'd Roman, and i'th'consul's view
> Slew three opposers. Tarquin's self he met,
> And struck him on the knee. In that day's feats,
> When he might act the woman in the scene,
> He prov'd best man i'th' field, and for his meed
> Was brow-bound with the oak. (2.2.91–98)

We may surmise from Volumnia's thanking the gods for Coriolanus's
wounds that Coriolanus had a less than ideal, less than warm, loving, or
nurturing childhood.

13. Greenblatt argues that Coriolanus's desire to "author…himself /…[as
if he] knew no other kin" (5.3.36–37) perfectly exemplifies the "will to
self-fashioning" (*Self-fashioning* 212) that, as Greenblatt sees it, is finally
"a choice…among possibilities whose range [is] strictly delineated by
the social and ideological system in force" (256).

14. To his own camp, he seems a god; to Aufidius's, "the devil" (1.10.16); to
both, something not quite human, something not quite natural, something a little bit monstrous.

15. Despite Menenius's calling the citizens "my countrymen" (1.1.55), there
is little reason to think that they are all men. It is not until act 2, scene
3 that we know for certain that the First Citizen is a man, the Second
Citizen calling him "sir" (2.3.3). The sexual ambiguity of citizens con-
tributes to the reader's sense of their ambiguous status as speaking subjects. How an audience receives this ambiguity, even though it is present
on the page, depends, of course, on the performance.

4 Pushing the Limits of Ecocriticism: Environment and Social Resistance in *2 Henry VI* and *2 Henry IV*

1. Patterson uses the term "tradition in the sense of something handed
down from the past" and "cultural" because "what was transmitted were
symbolic forms and signifying practices, a history from below encoded
in names and occasions, a memorial vocabulary and even a formal rhetoric" (38). As Stephen Greenblatt explains, "The unrest and class hostility

that afflicted England sporadically throughout Elizabeth's reign...led
to a series of disturbances that alarmed the propertied class...an official
concern sufficiently intense and wide-spread as to constitute something
like a national preoccupation" ("Murdering Peasants" 113).
2. In some ways, the relationship between York and Cade resembles that
between Prospero (Miranda, technically) and Caliban. Miranda teaches
Caliban language, which he then uses to curse his oppressors; York
plants the seeds of rebellion in Cade, but Cade is unable to mount
an effective rising, a credible rebellion, or a tenable subversive threat.
Still, the potential is as present with Cade as with Caliban, at least
theoretically.
3. Both Cade's ignorance and his stupidity are nicely captured in his comments about Lord Say. He argues thus:
       The Frenchmen
       are our enemies. Go to then, I ask but this: can he
       that speaks with the tongue of an enemy be a good
       counsellor, or no? (4.2.169–72)
The idiotic logic paints a picture of a questionable mind at work and an
equally unreliable group of followers. With such knowledge of interna-
tional politics and such skills at debate, Cade leads his "rebellion."
4. Although the king is clearly speaking in metaphor, his very powerful con-
demnation of the conversion of animals into meat does, to cite Dollimore
again, give "a voice, a part, a presence" (xxi) to a subverting of the enabling
speciesism that sponsors meat production and consumption. Speaking of
Gloucester's removal, the king asserts that
       as the butcher takes away the calf,
       And binds the wretch and beats it when it strays,
       Bearing it to the bloody slaughter-house,
       Even so remorseless have they borne him hence. (3.1.210–13).
5. Adams argues that both the metaphorical understanding of women
as consumable meat and the reification of animal flesh as masculating
food are fundamental to patriarchal thinking and to the joint oppression of women and animals. It is by no means anachronistic to apply
such ideas to the early modern period, since, as I have been showing throughout this book, early modern patriarchies conceptualize and deploy dynamically complementary power strategies against
women, people of color, dissidents, the natural world, "monsters," and
so on.
6. It is hard to agree with Wear on Tryon's fervency, since Tryon does, after
all, suggest kinds of meat to eat and times at which to eat them. Fervent
exponents would not do so. That he is, nevertheless, committed to an
argument against meat-eating, however, is clear enough. But there is
something else wrong with Wear's argument about Tryon being fervent:
the word "fervent" is belittling, dismissive, and ideologically loaded, in
much the way that words such as "shrill," "strident," and "shrieking"

are, each of which men use as belittling and dismissive descriptions of feminists and feminism. Tryon reasons soundly, not fervently, and though there are many points at which to query his logic (as we might do in arguing that there is a difference between faith and "fact," which itself is a difficult argument), we cannot reasonably call him fervent.

7. Tryon's logic is precise here. He argues against the idea that a vegetarian humanity would be destroyed by animals in the following way:

> There are no creatures that will hurt or be injurious to Mankind, but only Dogs, Bears, Foxes, and other of the like Nature, who are beasts of Prey, and fierce by kind, few or none of which are eaten. (285, T_7^R)

He argues that there are many animals that people do not kill or eat and that we are not overrun by them (308, X_2^V). What is particularly interesting is that this objection that we need to kill animals to avoid being overrun by them, like many of the claims against vegetarianism that Tryon counters, continues to arise from meat-eaters and meat industry people today in the twenty-first century.

8. Our distance from the reality of the foods we eat is greater now than in the early modern period, and the sanitized packages of little squares of flesh in the freezer and meat sections of supermarkets give us no sense of the animals from which the flesh came. If we are to eat flesh, such conceptual distancing is necessary. Tryon is trying to reduce the distance between the meal and the reality of its source. "Consider," he suggests, "how unpleasing it would be to most people, to behold the dead carcasses of beasts cut into pieces, and mangled, and all over bloody. And how nauseous and frightful a thing it would be to think of putting those begored Gobbits into our mouths, and feeding ourselves thereon...And how difficult a task it would be for many people...to kill the beasts for their own food" (305, X_1^R).

9. There has been very little written about early modern disease from an explicitly ecocritical perspective. The very best in this small group is Charles Whitney's article on plague pamphlets, entitled "Dekker's and Middleton's Plague Pamphlets as Environmental Literature."

10. When Margaret of Anjou in 2 Henry VI asks, "Why dost thou turn away and hide thy face? / I am no loathesome leper, look on me!" (3.2.74–75), she is speaking from the contemporary assumption that leprosy could be transmitted by sight, as well as by touch. See also Mullaney 34.

11. The view that illness was the direct result of spiritual defect and had nothing to do with the natural world was, in fact, a commonly held view. Andrew Boorde argues in 1547 that the first action a sick person should take to achieve well-being again is make confessions to his "goodly father" and "make his conscience clene" (Fol.V, B1^r). In Dearth's Death: Or, A Removall of Famine, William Gouge makes a similar socially detached evaluation of food shortages and the problems such shortages occasion and claims that shortages of food are divine punishments for

spiritual straying. He does not mention anything about full grain bins that could have been distributed. No matter how much Gouge may say that it is the sins of people—and he lists ten such sins (139–41) against the teachings of Scripture—it finally becomes a question of divine effect rather than one of causal relationships between people and the environment that, in his opinion, produces hunger and famine: "By want of corporall food God doth visibly demonstrate their folly in despising spiritual food" (141).

12. Cogan's recognition that fire is a preservation against the plague, though based on incorrect assumptions, remained a valid measure, whose effectiveness would eventually be factually understood and provable.

13. Falstaff's understanding of people is influenced both by Galenicism and by his beliefs about land husbandry: in making a case for sherry, he says that

> Hereof comes it that Prince Harry is valiant, for the
> cold blood he did naturally inherit of his father, he
> hath, like lean, sterile, and bare land, manur'd, hus-
> banded, and till'd with excellent endeavor of
> drinking good and good store of fertile sherris, that he
> is become very hot and valiant. (4.3.117–22)

14. Falstaff's comment resembles one Cade makes when he is offering to rule his followers. He promises to "burn all the records of the realm, my mouth shall be / the parliament of England" (4.8.14–15). Both men imagine a future in which they will have total power.

15. Falstaff becomes the epitome of infection, and in his dislocation from the centers of power, it is a telling comment on the play's conception of the natural world that by the end of the play he is, as Jeanne Addison Roberts so poignantly notes, "incurably tainted with gross animality" (89). What makes the insight of Roberts here so astute is that it links Falstaff, incurability, and animality. The ideological premise of 2 Henry IV is that the environment itself is irreducibly and incurably a pox on the face of all things civilized and a thing, therefore, to be feared. The further Falstaff moves from Hal's affections, the more he becomes a thing of the "wild."

5 Monstrosity in *Othello* and *Pericles*: Race, Gender, and Ecophobia

1. I have only been able to formulate these ideas because of the extraordinarily original and lucid arguments Carol J. Adams makes about "remembering" in her discussion of "Frankenstein's Vegetarian Monster" in her *The Sexual Politics of Meat*.

2. I identify Othello's speech here as the "here is my butt" speech with some humor, of course, and do not think or mean to imply that the word "butt" in Othello's speech denotes his buttocks. Nor do I mean to

suggest that Othello has a masochistic desire for punishment or that he wants to be beaten, whipped, or burned. Nevertheless, he *does* call attention to his body by repudiating it, and in his speech he participates in what Julia Kristeva might call his process of self-abjection. Paradoxically, his boastful grandiosity fails to glorify him; it defiles him. To borrow from Kristeva, we might say that through his speech, Othello is "abominated as ab-ject, as abjection, filth" (65).

3. Lynda Boose makes such an argument and claims that "Christian castration anxieties similar to those associated with Othello's threat of Otherness" (40) are "condensed" in Othello's final comments.

4. Certainly, the pictorial representations of "men whose heads do grow beneath their shoulders" show what appears to be at least a *kind* of headlessness.

5. The most vivid images of dismemberment, however, that we find in this play are Othello's raving threats of how he will discipline that supposedly hungry "white ewe" of his. At one point, he says, "I'll tear her all to pieces" (3.3.431), and at another, he screams, "I will chop her into messes...lest her body and her beauty unprovide my mind again" (4.1.200). It is her body that will unprovide him; it is her body that he disciplines. He sees her as an animal and treats her like an animal. One cannot help but think how like a dog Desdemona is to Othello, especially when he discusses her obedience:

> Sir, she can turn, and turn, and yet go on,
> And turn again; and she can weep, sir, weep;
> And she's obedient; as you say, obedient,
> Very obedient. (4.1.253–56)

6. Dawn H. Currie and Valerie Raoul are here summarizing Anthony Synnott's unpublished "The Two Bodies: The Social Construction of Self and Society."

7. It seems relevant here to mention that it is not only Desdemona who is strangled: but a kind of relationship is also strangled.

8. The play implicitly endorses Iago's claim that "these Moors are changeable / in their wills" (1.3.346–47). The play also endorses Iago's claim that "the Moor...will as tenderly be led by the nose / As asses are" (ll.399–402). Evans says that a supposed Moorish quality was "an astonishing credulity" (124), and Othello fits the stereotype perfectly.

9. In a post–O. J. Simpson double murder trial world, it is useful to observe the continuities between the early modern racist fascination with *Othello* and contemporary versions of such racism producing sensational media coverage and extraordinary media responses to the "trial of the century." Presentist criticism will also ask how this trial shapes our reading of the play.

10. The question here is one Terence Hawkes asks in regard to *The Tempest*, which he argues is "an arena for the sifting of the immense issue: what makes a man?" (29). In both plays, the question is contingent on an implicit and understood separation between "man" and "nature."

11. For a fuller treatment of this issue, see, for example, Kolodny, especially Chapter 1, 10–26; Stallybrass (1986); and Woodbridge, *The Scythe of Saturn*, especially Chapter 1, 45–85.
12. "America," by Jan van der Straet (see figure 5.2), similarly has cannibals as a feature virtually indistinguishable from the natural environment. In the background, animals are walking around nonchalantly amid the rich and luxuriant flora, while in a small and unobtrusive clearing, some equally nonchalant people are turning a leg of someone on a skewer over an open fire.
13. I am indebted to Professor Vernon McCarthy at Campion College (University of Regina) for bringing this article to my attention. Reported in the May 1, 1995, issue of *Time*, the source for much of the material seems to have come from an April 12, 1995, article in the *Hong Kong Eastern Express*. The text of this article may be found at <http://www.agathonvm.com/rlv/cr/eastexpr.html>, though the reliability of this source, given its conservative, antiabortionist, and religious affiliations is, for me, dubious.
14. In the summer of 2000, the Renault car company ran an advertisement that appeared on a billboard in downtown Auckland, New Zealand (and presumably in magazines and other places as well), with a picture of a car called "Scénic," the words beside it reading "Because Japanese cars all look the same." It is an image that I had wished to include in this book, but to each of a series of very persistent requests, the company repeatedly refused to give me permission to use the image. One can only speculate as to why the company refused to give permission, though this much is clear: the comment "they all look the same" (a comment with which the Renault advertisement bears a striking similarity) is one that rolls easily and frequently off of the tongue of racists. The xenophobic, anti-import sentiment of the advertisement is clear.
15. See, for instance, Maggie Kilgour's work, which looks at the symbolic cannibalism of communion.
16. For discussions about relationships between colonialist discourse and the discourse of cannibalism, see Philip P. Boucher, Peter Hulme (1978), Richard B. Moore, and Stephen Orgel. M. E. Montaigne offers an early modern "documentary" statement about cannibalism. William Arens argues that the discursive emphasis on cannibalism in European descriptions of New World cultures (while other elements of those cultures go unremarked) is perhaps more descriptive of the Old World than the new.

We should also note here that some of the most recent work in ecocriticism has developed very productive connections between postcolonial theory and ecocriticism. This work argues, to some extent, that discussions about human freedom need to be conducted within frameworks that recognize and address hierarchies between human and nonhuman communities and that

postcolonial ecocriticism—like several other modes of ecocriticism—performs an *advocacy* function both in relation to the real

world(s) it inhabits and to the imaginary spaces it opens up for con-
templation of how the real world might be transformed. (Huggan
and Tiffin 13)

17. Frederick Kiefer is one critic who has explicitly addressed "nature" issues
in the play but in thematic terms that fail to recognize the ideological
significance of the topics under discussion. The consequence is that he
obscures the didactic potentials of the literature by unwittingly repro-
ducing the very structures he critiques and that a more rigorously politicized reading would not reproduce. He discourses learnedly on how a
culture such as Shakespeare's "can personify nature as a woman" (209)
but completely undercuts the significance of what is potentially a radical materialist analysis when he duplicates the gendering of nature in
such phrases as "nature herself," "her purposes," what "she is capable
of" (212), and so on. Surely there is something wrong-headed about
ascribing consciousness and gender to nature. Surely the very fact that
women and nature in *Pericles* are so closely linked is important because
the denial of voice and agency to each is the precondition of their subju-
gation to men.

Moreover, in a world (such as Shakespeare's or ours) where women
are butchered, where violence against women remains a problem in
every corner, the symbolic importance of metaphors linking sexuality
with food and the environment seems relatively immaterial. What seems
more important to praxis is to understand how and why the metaphors
work, some of which I discuss further below.

18. Each of these writers (and there are scores more) describe very much the
same interstitiality of monsters, but each uses slightly different terms.
Even with such a small group, we have gotten such mixed terms as "monsters," "prodigies," "hybrids," and "abjects." Paré's definitional project,
it would seem, is an ongoing one, terminology still being a problem.

19. See also figure 5.1.

6 DISGUST, METAPHOR, WOMEN: ECOPHOBIC CONFLUENCES

1. Greenblatt's thesis is a bit problematic, since in the drawing of any
boundary, as Derek Gregory has argued the Derridean point, "each side
folds over and implicates the other in its constitution" (Gregory 72).
John Carlos Rowe argues much the same position, claiming that "the
'margin' is always already constituted by its exclusion, by a powerful
act of cultural repression" (155). Peter Hulme argues along the same
deconstructive lines in pointing out that the "boundaries of community
are often created by accusing those outside the boundary of the very
practice on which the integrity of that community is founded" (*Colonial
Encounters* 85). The Other can never really, then, be absolutely outside
of or Other to the discursive system that seeks to produce it. Richard

Bernstein puts it best: "At the heart of what we take to be familiar, native, at home—where we think we can find our center—lurk (is concealed and repressed) what is unfamiliar, strange and uncanny" (174).

2. Weeds, of course, are a form of natural pollution (essentially, rot) in this play, and they are abundant. They define nature in terms heavily inscribed with human investments. Imagined as having no practical value and as being detrimental to things that do have practical value, weeds in this play devalue all that they are associated with. They epitomize amoral luxuriance and anthropomorphize nature as corrupt and rotten. They cross boundaries drawn for utilitarian purposes, and they are a threat because their order stands in defiance and challenge to human order. This is not to say that weeds were uniformly loathed in the early modern period. Indeed, a weed such as dandelion (also known as lion's tooth, bitterwort, wild endive, priest's crown, piss-a-bed, Irish daisy, blow ball, yellow gowan, puffball, clock flower, swine snout, fortune-teller, and cankerwort) is a frequent guest in many herbals of the time which were commonly available. Among these are the *Book of Soveraigne Approved Medicines and Remedies* (1577), a catalog of medicines and methods for their preparation approved by the government and readily available in England; William Turner's *A New Herball* (1551); Banckes' *An Herball* (1525); Sir Thomas Elyot's 1541 *Castel of Helth;* Thomas Moulton's 1540 *This is the myrour or glasse of helthe,* among many published works detailing common medical practices with what are in other forum called weeds.

3. Correspondences between the natural and social worlds are indeed abundant in Shakespeare, from the disorderly and carnivalesque world of the witches in *Macbeth*, with their "fog and filthy air" (1.1.11), to the proclamations of Ulysses in *Troilus and Cressida*, where we are given the rhetorical question asking

> What raging of the sea, shaking of earth,
> Commotion in the winds, frights, changes, horrors,
> Divert and crack, rend and deracinate
> The unity and married calm of states
> Quite from their fixture? (1.3.97–101)

In his next breath, Ulysses gives the answer: "O, when this degree is shaked, / Which is the ladder of all high designs, / The enterprise is sick" (ll.101–3). Disorder in the natural world is disorder in the social world.

4. The enseamed bed here is Gertrude's, and it is eminently disgusting, rotten, and dirty to Hamlet because it flies in the face of the kind of order that Hamlet would have wished to have seen maintained.

5. Leonard Tennenhouse's discussion of women as perceived sources of pollution is intriguing, especially for its treatment of issues about dismemberment as enactments of disciplinary responses to perceived or constructed sources of pollution. The analysis, however, is anthropocentric and attempts no comment about the significance of the ways in which the environment is conceptualized, implicitly or explicitly, in the early modern metaphors of pollution.

NOTES   145

6. For an extensive analysis of the history of menstruation as it is discursively produced in (for the most part Western) history, see Janice Delaney, Lupton, and Toth.
7. See Sturgeon, *Ecofeminist Natures*, 167–96.
8. A problem with the concept of the "literal" is that it assumes an unproblematic relation exists between words and what they "mean." It is a problem that John Searle adds to in "Metaphor" when he presumes that there are particular meanings that are firm, true, solid, and unquestionable; that definitions are facts and are indisputable (like the pain that comes from a sudden unanaesthetized amputation), facts from which metaphors deviate and are, as Samuel Levin describes Searle's notion, "defective" (112). Jonathan Culler argues that an unproblematized acceptance of the validity of the literal/figurative split is naïve. He maintains that "the literal is the opposite of the figurative, but a literal expression is also a metaphor whose figurality has been forgotten" (148). Culler goes on to explain that "the distinction between the literal and the figurative [is] essential to discussions of the functioning of language" and that rather than "treating figures as deviations from proper, normal literality," we have to, he reminds us, remember that all so-called literal language is forgotten figurality (150).
   Historically, we have not looked at language the way Culler does. As Raymond W. Gibbs explains, it is a "centuries-old belief that literal language is a veridical reflection of thought and the external world whereas figurative language distorts reality" (254).
9. I. A. Richards argues that "metaphor is the omnipresent principle of language" (92). Max Black talks about "the ubiquity of metaphor" and asks why it is such a baffling issue (21). Paul de Man argues on the basis of his discussion of John Locke, Etienne Bonnot de Condillac, and Immanuel Kant that since language uses metaphor, no discussions of metaphor can logically maintain a distinction between the literal and metaphorical, that "it turns out to be impossible to maintain a clear line of distinction between rhetoric, abstraction, symbol, and all other forms of language" (26).
10. Black makes a very similar argument from the same play (21).
11. For my purposes, Lakoff and Johnson's very simple working definition of metaphor is sufficient: in *Metaphors We Live By*, they define metaphor as "the understanding and experiencing [of] one kind of thing in terms of another" (5). It is a fairly commonplace definition. Max Black similarly proposes that "every implication-complex supported by a metaphor's secondary subject...is a *model* of the ascriptions imputed to the primary subject: Every metaphor is the tip of a submerged model" (30) and that metaphor is "an instrument for drawing implications grounded in perceived analogies of structure between two subjects belonging to different domains" (31).
12. Paul de Man, too, seems inclined to this view when he ponders the relationship between cognition and metaphors: "One may wonder," he

explains, "whether the metaphors illustrate a cognition or if the cognition is not perhaps shaped by the metaphors" (14).

13. Annette Kolodny's *The Lay of the Land* was a pioneer among analyses of metaphors in literature that conflate women and land.

14. Ted Cohen, though he is not among those Tilley cites, argues this position well. "There is," he maintains,

> a unique way in which the maker and the appreciator of a metaphor are drawn closer to one another. Three aspects are involved: (1) the speaker issues a kind of concealed invitation; (2) the hearer expends a special effort to accept the invitation; and (3) this transaction constitutes the acknowledgement of a community. All three are involved in any communication, but in ordinary literal discourse their involvement is so pervasive and routine that they go unmarked. The use of metaphor throws them into relief, and there is a point in that. (6)

Unlike in "ordinary literal discourse," in metaphorically intended discourse, Cohen argues, we see a move to "initiate explicitly the cooperative act of comprehension which is, in any view, something more than a routine act of understanding" (7).

15. See Ken Hiltner's "Renaissance Literature and Our Contemporary Attitude toward Global Warming" for a detailed and illuminating discussion of seventeenth-century London's air pollution problem.

16. It also seems a mistake to argue, as does Keith Thomas, that the "continuing use of animal analogy and metaphor in daily speech reinforced the feeling that men and beasts inhabited the same moral universe" (99). Such an argument does not square well with the changing early modern uses and conceptions of the natural world and of animals, which Thomas himself describes. Increasingly, the view of "nature as machine" replaced ideas about "nature as organism" in the early modern period. The "transition [is] from the organism to the machine as the dominant metaphor binding together the cosmos, society, and the self into a single cultural reality—a world view" (Merchant xxii). Thomas's argument would suggest that machines belong to the same moral universe as people—which they do not.

17. The metaphor I use here of organicism as a kind of childhood of mechanism seems supported by Merchant's implicit allusion to Peter Laslett's book about childhood, the title of which (*The World We Have Lost*) provides the opening words for her book.

18. Howard Felperin seems inclined to argue that Hermione is "tongue-tied" and that her contorted and tortuous syntax perhaps partly justifies the wild imaginings of Leontes. But it is more productive for my purposes to look not at how her words might damn her, but at the ideological effects of her silence, at the workings of the words that are inscribed in the space left empty by her silence. What we are presented with is not merely a silencing, though, nor even an erasure, but an ossification, a pause on the DVD of Hermione's life, a pause held for Leontes

to work out his matters. Hermione, a very real material presence, must, in this play, be denied her material realities for the man whose matters weigh more heavily in the sexist scales that the play presents. Hermione's presence—like a video—can be turned on or off, depending on what the matters demand in the male arena that views and controls her. Such is her dramatic function, and it is one that is startlingly similar to the dramatic function of the bear. When it is needed, it is called in, and it is abandoned just as easily.

19. *The Winter's Tale* might be viewed as radically subversive because it dramatizes an open and intelligent revolt against authority from the mouth of a woman—Paulina. And like a scared child wielding a loaded gun at a threatening felon, Leontes unsurprisingly fixes on "force" (2.3.62)—his only sure weapon against Paulina's words. He bellows in an uncontrolled rage that he will have her burnt (2.3.114). The startling juxtapositioning of his unwieldy rage and irrationality with her relative calm and intelligence is all the more disruptive and subversive of order in his world. But, for all of this, the play ultimately contains any truly radical position that it gestures toward. All is forgiven in the end, and the order that allows the suffering in the play is completely intact.

20. Laird's purpose is more to look at the relationship between language and perceived realities in *The Winter's Tale*. He explains that for Leontes, "The aspect of reality which inspires his mistrust is the indeterminacy of things, their shifting shapes and meanings which he is resolved to fix and stabilize" (31), and that it is Hermione's words that elicit this mistrust.

21. With few exceptions, the thinking of the time posits women's silence within an essentialist ecological framework. Henry Cornelius Agrippa, though certainly not free from his own versions of essentialism and paternalism, is atypical in arguing against the view that women's silence is part of the natural order, that it is more "by force of armes [that women] are constrained to giue place to men, and to obeye theyr subdewers, not by no naturall, no diuyne necessitie or reason, but by custome, education, fortune, and a certayne Tyrannicall occasion" (Sig G, G^V).

22. When we are introduced to Prospero, he is almost paranoically concerned with being heard, repeatedly asking if Miranda is listening: "Dost thou attend me" (1.2.78); "Thou attendst not" (l.87); "Dost thou hear?" (l.106). His power is in language, and without his books, he is but a sot.

23. It is not just the play's "commentary on the limits of human knowledge or power" about which Helena Feder speaks in a brilliant rebuttal to Frederick W. Turners's "Cultivating the American Garden" (Feder 43) that reveals ecophobia, though this is very significant in itself. As I have repeatedly argued throughout this book, any anxieties about loss of control to nature are invariably ecophobic; but there is more here worthy of comment. The hapless Antigonus—in his radically understated unfortunate and fatal encounter with a bear, which "tore out his shoulder

bone" (3.3.95) and ate him—is subject to an unpredictable and, in many ways, inexplicable nature. Critics have long sought to explain away (with little success) "Exit, pursued by a bear." The fact that this is a nature that the text characterizes as sentient makes the whole situation much more frightening, the whole of nature much more sinister and hideous. Moments before the demise of Antigonus, "the sea mock'd" (3.3.99), "swallow'd with yest and froth" (3.3.93) the unfortunate mariners who accompanied Antigonus. The anthropocentric image is of the environment as some kind of disaffected subject (in competition with the men), whose *raison d'être* is to cause chaos, pain, suffering, or loss—rather like a computer software specialist who designs viruses. It is ruthless, both in the anthropocentric language that the characters in the play use to describe it, and in the audience's understanding of it as a hostile threat to order and goodness. This is perhaps not so surprising, since Judeo-Christian society has a long history of allegorizing the environment—one has only to think of the tree that bears the fruit that yields knowledge of good and evil.

24. Dorothea Kehler argues this position in "Teaching the Slandered Women of *Cymbeline* and *The Winter's Tale*." Drawing heavily on the work of Jean Baudrillard, she argues that depictions of women in *The Winter's Tale* follow Baudrillard's concept of *simulacra*, models "without origin or reality" (Baudrillard 1).

25. We do not see disaster of the kind that we see, for instance, in other plays of Shakespeare where there is similar substantial boundary transgression—*Othello*, *Titus Andronicus*, and even *Romeo and Juliet* come to mind (the latter because the warring families could be argued to constitute a version of class conflict and can unquestionably be said to profile a forbidden interbreeding).

26. Often metaphorical, pollution in the play covers a broad field: epistemological pollution—rotten opinions (2.3.90–91) and "infected knowledge" (2.1.41–42); gender pollution—the blurred gender boundaries of the "mankind witch" (2.3.68); sanguinary pollution—Polixenes's infected blood (2.1.57–58), "an infected jelly" (1.2.417–18); and air pollution—the infected air of Sicilia (5.1.168–70). The image of environmental pollution working allegorically as a metaphor for the pollution of the body politic, in the infected air, is common. There are times when we really do not like the environment that this play describes, and the two dozen references, oblique and direct, to pollution in *The Winter's Tale* contribute to this ecophobic reaction. By far the most important kind of pollution in *The Winter's Tale* is perhaps best described as "genetic." It is on this string that most of the plays thematic issues hang. It reaches its acme in the play's pastoral interlude. It is a formal debate between Polixenes and Perdita on the division between Art and Nature resting on anxieties first about crossbreeding and second about definitions, classifications, and naming.

27. Indeed, it is not only flowers, trees, and other plants but, particularly, livestock and domestic animals that are being bred by this time.

28. I am indebted to Greta Gaard (Personal communication, February 3, 2010) for bringing this article to my attention. It is, as far as I know, the only published material explicitly linking ecofeminist and ecocritical matters with theories about disgust.

7. Staging Exotica and Ecophobia

1. Disguised as "Sir Topas the curate" (4.2.21), the Clown enters to enact an exorcism. He commands the "hyperbolical fiend" (l.25) to get out of Malvolio. All that the poor Malvolio can do is protest his sanity, which he does repeatedly (l.29, l.40, ll.47–48, l.88, l.106–7, l.115).
2. I discuss the "exorcism scene" of *Lear* in chapter two above.
3. Richard Kerridge argues persuasively that "environmentalism has a political weakness in comparison with feminism: it is much harder for environmentalists to make the connection between global threats and individual lives. Green politics cannot easily be, like feminism, a politics of personal liberation and empowerment" (6). Kerridge is not arguing that such connections are absent but only that people may often fail to see them.
4. From a bioregional perspective, thinking locally as Nicholas Culpeper, William Harrison, and other early moderns seem to do is ecologically sound and desirable, but the motives are national, not ecological. It makes a difference.
5. The love that dare not speak its name, or what Burton calls "a nastiness and abomination even to speak of" (653), is spoken of here—but not in English. The section was originally set off in Latin. Winfried Schleiner has argued that Burton's use of Latin for a disquisition about sodomy becomes particularly important when we consider that "few women of the period would have had the Latin to read it," and he suggests that "a twelve-year-old grammar school student in the early seventeenth century [would] have [had] enough Latin to read" it (173). Writing in Latin is not, per se, *praeteritio*, though Schleiner seems to be arguing that it is. Moreover, there is, in fact, very little "passing over" or "not saying" in this chapter, and Schleiner's argument seems a bit muddled. When Burton says "I will pass over X," his mention of X is clearly not a passing over. If a person says "I will refrain from telling you that it is raining," then "you" *have* been told.
6. By February 1973, "homosexuality" and madness are still linked:

    We are told, from the time that we first recognize our homosexual feelings, that our love for other human beings is sick, childish, and subject to "cure." We are told that we are emotional cripples, forever condemned to an emotional status below that of the "whole" people who run the world. The result of this in many cases is to contribute to a self-image that often lowers the sights we set for ourselves in life, and many of us asked ourselves, "How could anybody love me?" or "How can I love somebody who must be just as sick as I am?" (reproduced in Bayer 119, from a memo to the American Psychiatric Association)

Until December 15, 1973, *The Diagnostic and Statistical Manual of Mental Disorders* (*DSM*)—the bible of definitions for the American Psychiatric Association—diagnosed homosexuality as a "sociopathic personality disturbance" (Goss 45). Any act "not usually associated with coitus" (*DSM-II* 44) qualified (or could qualify) as a "sexual deviation" and, ipso facto, as a "mental disorder."

Changes in the *DSM* (specifically the removal of homosexuality as a mental disorder), however, do not necessarily mean changes in the perceptions about gays, lesbians, bisexuals, and transgender people. Even as recently as June 12, 1995, the ultraconservative Canadian *Alberta Report*, in two overtly homophobic articles about gay rights, dig up a psychologist who believes that "homosexuality is a personality disorder and a symptom of a fragile identity" (Sillars 31) and a sociologist who argues about the "pathological nature of homosexuality" (Woodard 30).

7. The word "spirit" here is interesting, and it has been observed that Jorden is the first theorist to locate the brain as well as the uterus as a source of hysteria (see also Veith 122–23 and Neely, "Documents" 320, n16).

8. See also Peter Stallybrass, "*Macbeth* and Witchcraft" and Stuart Clark, "Inversion, Misrule, and the Meaning of Witchcraft."

9. In *The Changeling*, Jasperino makes reference to the superstition that witches could control weather when he tells Alsemero that, even if he "could buy a gale amongst the witches" (1.1.17), he could not hope for better winds than were prevailing.

10. Lady Macbeth's supplication for "thick blood" resonates as a call for madness: as Babb explains, "thick blood is melancholic blood" (84).

11. Leiss, Kline, and Jhally review of Marx's discussion about the growth of commodity fetishism —see also Karl Marx 43–144.

12. There are many more characters we can discuss, but these five serve our illustrative purposes fairly well.

13. The empty vessel is the disempowered one, and when Anthony Nixon explains that "[Man]...as an emptie ship is to be fenced and furnished with conuenient tackling: So is a Man's life with the effect of Prudence" (125, R_3^R), Mankind (generically, not males anatomically) is the passive subject. Men tend not to be empty vessels unless they take the whole category of humanity with them.

14. This is, of course, part of the racist and speciesist economy through which, as Cary Wolfe has argued, commodification and exploitation take form:

The humanist concept of subjectivity is inseparable from the discourse and *institution* of a speciesism which relies on the tacit acceptance that the full transcendence of the human requires the sacrifice of the "animal" and the animalistic, which in turn makes possible a symbolic economy in which we can engage in a "non-criminal

putting to death" (as Derrida phrases it), not only of animals but of
humans as well by marking *them* as animal. (Wolfe 43)

15. Shakespeare does not idealize this precontact world as one without conflict. Ariel and his arboreal confinement is a case in point.

16. Although most audiences will probably remember Portia or Shylock
before Antonio, it is important to remember from whom the play takes
its title: Antonio is the "Merchant of Venice." He is *central*, at least symbolically, to this play. As Barbara Babcock has argued, "What is socially
peripheral is often symbolically central" (32).

17. John Boswell, although he tends to read the past myopically in terms
of the present, has noted historical similarities between the treatments
of Jews and sexual minorities: as he explains in *Christianity, Social
Tolerance, and Homosexuality,*

> The fate of Jews and gay people has been almost identical through-
> out European history... [and] the same laws which oppressed
> Jews oppressed gay people; the same groups bent on eliminat-
> ing Jews tried to wipe out homosexuality; the same periods of
> European history which could not make room for Jewish distinc-
> tiveness reacted violently against sexual nonconformity... and
> even the same methods of propaganda were used against Jews and
> gay people—picturing them as animals bent on the destruction of
> the children of the majority. (15–16)

So close are the terms through which anti-Semitism and discourses
of sodomy define themselves that in early modern France, Christian/
Jewish mixing could be punished as sodomy. E. P. Evans relates through
secondary sources that "a certain Johannes Alardus or Jean Alard, who
kept a Jewess in his house in Paris and had several children by her... was
convicted of sodomy on account of this relation and burned, together
with his paramour, 'since coition with a Jewess is precisely the same as
if a man should copulate with a dog'" (Evans 153; Döpleri 157 cited by
Evans). Not only do we have a confluence of discourses of sodomy and
anti-Semitism here, but a mixing of both with thinking about bestiality. In addition, we must also wonder about the diminutive ending on
the word "Jew," an ending also used on the word "Negro." Neither the
word "Christian" nor "white" have such endings to denote the female
form. In thinking about Christianess and whitess, it is clear that sexism
is also writ large in the manner of the accounts of both Evans and his
sources.

18. Derek Gregory succinctly notes that "it is a commonplace of feminist
history that 'Nature' has been coded feminine within the Western intellectual tradition" (129).

19. A 1993 case of the Supreme Court of Canada (*Regina v. Litchfield*,
November 18, 1993) suggests that it is the duty of a woman to offer
her body for anatomization by the judicial system. This anatomization would ostensibly allow prosecution against the accused,

Dr. Bryant Floyd Litchfield, charged with fourteen counts of sexual assault. The argument that the chambers judge (Justice McDonald) makes, in short, is that "the law should be interpreted in a fashion that would parcel women into body parts" (I am indebted to Michigan lawyer R. Pilling-Johnson for succinctly articulating the terms of the law to me). As recorded in the court transcripts, "The chambers judge... ordered that there be three different trials depending on the part of the complainants' body that was involved in the assault. Thus, there was to be one trial for allegations involving the complainants' genitalia, a second trial for allegations concerning the complainants' breasts, and a third trial for any other matters" (*Regina v. Litchfield* 97h). One might pause to note here Rudolf Wittkower's observation that in the early modern period, scientists would group monstrosities "according to the parts in which the irregularity occurs. Head, breast, arms, hands, etc., follow each other in long-winded chapters" (Wittkower 189).

20. Portia explains that

    It is enacted in the laws of Venice,
    If it be proved against an alien
    That by direct or indirect attempts
    He seeks the life of any citizen,
    The party 'gainst which he doth contrive
    Shall seize half his goods; the other half
    Comes to the privy coffer of the state;
    And the offender's life lies in the mercy
    Of the Duke. (4.1.347–56)

    It is *because* Shylock is an alien that he is treated with such contemptuous discrimination.

21. Bassanio is unlike the other suitors (among them, a "tawny Moor" [2.1.SD]—who, in the stunningly racist economy of the text, is the dullest of the suitors and therefore chooses gold—and the Prince of Arragon, apparently not black, but a foreigner nonetheless whose intellect allows him only to choose wrongly the silver casket). Bassanio is a good homegrown sort of fellow, and, yes, he correctly chooses lead, the *summum bonum* in the inverted hierarchy. Despite appearances, though, Bassanio *does* choose gold. He chooses Portia, whose "sunny locks / Hang on her temples like a golden fleece" (1.1.169-71)—and, of course, it is her gold that he is after. While Portia actively resists her role as passive object waiting to be chosen, the scripted ideal remains essentially intact. The sense at the end of the play is that her transgressions will be *forgiven* and that she will behave herself according to the ideals that she has so clearly challenged.

22. For an in-depth discussion about this issue, see David George Hale.

23. In the famous "Ditchley Portrait" (artist unknown, 1592), Elizabeth appears on top of a map of England; in *Sphaera Civitatis* (John Case, 1588), Elizabeth embraces "a pre-Copernican universe" (Strong 126);

in "Europa" (artist unknown, 1598), Elizabeth appears *as* a map. As Roy Strong explains,

Her right arm is made up of Italy; her left of England and Scotland, her feet are planted in Poland. To the left of her, the Armada is defeated; to the right, a triple-headed Pope rides away in a boat rowed by clergy and escorted by a fleet of ships, all of which are numbered and allude to papal allies. (114)

24. See Nancy Vickers (1981, 1985, and 1986) for discussions about the *blazon* in terms of questions about discursive dismemberment.

25. Metaphorical equations between rape and exploitation constitute a huge and sprawling topic, only a fraction of which I am concerned with here. Numerous early modern texts equate military conquest with rape— *Titus Andronicus*, *Lucrece*, and *Henry V* are just a few—but this topic has been discussed extensively by other critics. See, for example, Linda Woodbridge (*Scythe*) and Heather Dubrow, especially Chapter two. My concern is less with military than environmental invasion. Relationships among issues of rape and men's control of property undergird Marion Wynne-Davies's discussions about *Titus Andronicus*. Susan Brownmiller also addresses such relationships. The mentality that sees women as environmental commodities is one that does not blanch at prospects of violence—either to women or the environment.

26. For each item in this list, Woodbridge provides, in footnotes, extensive examples from Elizabethan texts (see *Women and the English Renaissance* 268–70).

27. I am building, in part, on Elaine Scarry's argument that "to have no body is to have no limits on one's extension out into the world" (207); but we are clearly arguing in quite opposite directions. Scarry is arguing that to have no body is empowering, an argument similar to one Jane Gallop makes, where, "by giving up their bodies, men gain power—the power to theorize, to represent themselves, to exchange women" (67). Such a contention, however, assumes that discursive emphasis on the body results in what Scarry calls "intensely embodied" (207) figures. I argue that such is not the case; I have yet to see empowering representations of figures whose bodies have been blurred, dismembered, or consumed.

28. This particular phrase occurs in *Hamlet* (1.1.137). We also find the word "womb" in *The Winter's Tale*, where it is used as a verb in association with the earth: the "earth wombs" (4.4.483), or encloses, in Florizel's usage.

29. Annette Kolodny offers one of the early ecofeminist discussions of the relationship between representations of women and the New World environment. Kolodny maintains that the experience of the American landscape "is variously expressed through an entire range of images, each of which details one of the many elements of that experience, including eroticism, penetration, raping, embrace, enclosure, and nurture, to cite only a few" (150). Her discussions about the representation and

conceptualization of the New World in feminine terms remain valid. An enormous and growing body of work from the ecofeminist community further develops theories about the conceptual links between representations of women and the environment.

8   THE ECOCRITICAL UNCONSCIOUS: EARLY MODERN SLEEP AS "GO-BETWEEN"

1. My intention here, however, is not to confirm and endorse Agamben's rather naïve and simplistic idea that *"Dasein* is simply an animal that has learned to become bored; it has awakened *from* its captivation *to* its own captivation" (2004: 70); instead, without fully endorsing Agamben, we can understand that his continuing contribution to the ongoing discussion about "the animal question" is one of a movement away from the separation of human and animal.

WORKS CITED

Abate, Corinne S., and Elizabeth Mazzola. "Looking for Goneril and Regan." *Privacy, Domesticity, and Women in Early Modern England.* Ed. Cristina León Alfar. Aldershot, England: Ashgate; 2003. 167–98.

Adams, Carol J. *The Sexual Politics of Meat: A Feminist-Vegetarian Critical Theory.* New York: Continuum, 1991.

Adelman, Janet. *Suffocating Mothers: Fantasies of Maternal Origin in Shakespeare's Plays,* Hamlet *to* The Tempest. New York: Routledge, 1992.

Agamben, Georgio. *The Open: Man and Animal.* Ed. Werner Hamacher. Stanford, CA; Stanford UP, 2004.

Agrippa von Nettesheim, Heinrich Cornelius. *A Treatise of the Nobilitie and excellencye of woman kynde.* Trans. David Clapham. London: Thomæ Bertheleti, 1542. STC 203.

Albion, Robert Greenhalgh. *Forests and Sea Power: The Timber Problem of the Royal Navy, 1652–1862.* Cambridge, MA: Harvard UP, 1926.

Alfar, Cristina León. *Fantasies of Female Evil: The Dynamics of Gender and Power in Shakespearean Tragedy.* Newark: U of Delaware P, 2003.

Anonymous. *A booke of soueraigne approued medicines and remedies as well for sundry diseases within the body as also for all sores, woundes,...Not onely very necessary and profitable, but also commodious for all suche as shall vouchsafe to practise and vse the same.* London: Thomas Dauson, and Thomas Gardyner, Nouembre 24. 1577. STC 309:07.

Arens, William. "Rethinking Anthropology." *Cannibalism and the Colonial World.* Ed. Francis Barker, Peter Hulme, and Margaret Iversen. Cambridge, Cambridge UP, 1998. 39–62.

Arnold, Jean. "Letter," "Forum on Literatures of the Environment," *PMLA* 114.5 (October 1999): 1089–90.

Aubrey, James R. "Race and the Spectacle of the Monstrous in Othello." *CLIO: An Interdisciplinary Journal of Literature, History, and the Philosophy of History* 22.3 (Spring 1993): 221–38.

Azzarello, Robert. *CFP: Queer Ecocriticism and Theory.* June 17, 2008. <http://cfp.english.upenn.edu/archive/Theory/1439.html> July 1, 2008.

Babcock, Barbara. *The Reversible World: Symbolic Inversion in Art and Society.* Ithaca, NY and London: Cornell UP, 1978.

Bacon, Francis. *The Essays.* Ed. John Pitcher. New York: Penguin, 1986.

Bankes (anonymous). *Here begynnyth a newe mater, the whiche sheweth and treateth of ye vertues [and] proprytes of herbes, the whiche is called an herball Cum gratia [and] priuilegio a rege indulto.* London: Rycharde Banckes, 1525. STC (2nd ed.) 13175.1.

Baudrillard, Jean. *Simulacra and Simulation.* Trans. Sheila Faria Glaser. Ann Arbor: U of Michigan P, 1994.

Bayer, Ronald. *Homosexuality and American Psychiatry: The Politics of Diagnosis.* New York: Basic Books, 1981.

Behringer, Wolfgang. "Climatic Change and Witch-Hunting." *Climatic Change* 43.1 (1999): 335–51.

Belsey, Catherine. *Why Shakespeare?* New York: Palgrave Macmillan, 2007.

Berger, Harry. *Second World and Green World: Studies in Renaissance Fiction-Making.* Berkeley and Los Angeles: U California P, 1988.

Bernstein, Richard. *The New Constellation: The Ethical-Political Horizons of Modernity/Postmodernity.* Cambridge, MA: MIT Press, 1992.

Bevington, David. "Asleep Onstage." *From Page to Performance: Essays in Early English Drama.* Ed. John A. Alford. East Lansing: Michigan State UP, 1995. 51–83.

Black, Max. "More about Metaphor." *Metaphor and Thought, 2nd Edition.* Ed. Andrew Ortony. Cambridge: Cambridge UP, 1993. 19–41.

Boehrer, Bruce. *Shakespeare among the Animals: Nature and Society in the Drama of Early Modern England.* New York: Palgrave Macmillan, 2002.

Boorde, Andrew. *The Breuiary of Helthe.* (Facsim edition of 1547). Amsterdam and New York: Da Capo Press, 1971.

———. *The Fyrst Boke of the Introduction of Knowledge. A Compendyous Regyment; or, A Dyetary of Helth Made in Mountpyllier. Barnes in the Defence of the Berde (1562).* (Facscimile reproduction). London: Adamant Media, 2001.

Boose, Lynda E. "'The Getting of a Lawful Race': Racial Discourse in Early Modern England and the Unrepresentable Black Woman." *Women, "Race," and Writing in the Early Modern Period.* Ed. Margo Hendricks and Patricia Parker. London and New York: Routledge, 1994. 35–54.

Borlik, Todd Andrew. "Mute Timber?: Fiscal Forestry and Environmental Stichomythia in the Old Arcadia." *Early Modern Ecostudies: From the Florentine Codex to Shakespeare.* Ed. Ivo Kamps, Thomas Hallock, and Karen Raber. New York: Palgrave Macmillan, 2008.31–53.

Boswell, John. *Christianity, Social Tolerance, and Homosexuality: Gay People in Western Europe from the Beginning of the Christian Era to the Fourteenth Century.* Chicago and London: U. of Chicago P, 1980.

Boucher, Philip P. *Cannibal Encounters: Europeans and Island Caribs, 1492–1763.* Baltimore and London: Johns Hopkins UP, 1992.

Bownd, Nicholas. *Medicines for the plague: that is, godly and fruitfull sermons vpon part of the twentieth Psalme, full of instructions and comfort: very fit generally for all times of affliction, but more particularly applied to this late visitation of the plague*. London: Adam Islip for Cuthbert Burbie, 1604. STC 3439.

Bray, Alan. *Homosexuality in Renaissance England*. London: Gay Men's Press, 1982.

Bristol, Michael D. "Lenten Butchery: Legitimation Crisis in *Coriolanus*." *Shakespeare Reproduced: The Text in History and Ideology*. Ed. Jean E. Howard and Marion F. O'Connor. Methuen: New York, 1987. 207–24.

Brown, Georgia. "Defining Nature through Monstrosity in *Othello* and *Macbeth*." *Early Modern Ecostudies: From the Florentine Codex to Shakespeare*. Ed. Ivo Kamps, Thomas Hallock, and Karen Raber. New York: Palgrave Macmillan, 2008. 55–76.

Brownmiller, Susan. *Against Our Will: Men, Women and Rape*. New York: Bantam, 1975.

Bruckner, Lynne, and Dan Brayton (eds.). *Ecocritical Shakespeare*. Aldershot, England: Ashgate, 2011.

Buell, Lawrence. *The Future of Environmental Criticism: Environmental Crisis and Literary Imagination*. Oxford: Blackwell, 2005.

Burke, Kenneth. "*Coriolanus* and the Delights of Faction." *Hudson Review* 19.2 (Summer 1966): 185—202; rpt. in *Kenneth Burke on Shakespeare* (Ed. Scott L. Newstock). West LaFayette, IN: Parlor Press, 2007. 129–49.

Burton, Robert. *The Anatomy of Melancholy*. Ed. Floyd Dell and Paul Jordan Smith. New York: Tudor Publishing, 1960.

Butler, Judith. *Bodies that Matter: On the Discursive Limits of "Sex."* New York and London: Routledge, 1993.

Cartelli, Thomas. "Jack Cade in the Garden: Class Consciousness and Class Conflict in *2 Henry VI*." *Enclosure Acts: Sexuality, Property, and Culture in Early Modern England*. Ed. Richard Burt and John Michael Archer. Ithaca, NY: Cornell UP, 1994. 48–67.

Case, John. *Sphaera civitatis, authore Magistro Iohanne Caso Oxoniensi, olim Collegii Diui Iohannis Praecursoris socio*. Oxoniae: Excudebat Iosephus Barnesius, 1588. STC 189:02.

Chojnacki, Stanley. *Women and Men in Renaissance Venice: Twelve Essays on Patrician Society*. Baltimore: Johns Hopkins UP, 2000.

Churche, Rooke. *An olde thrift newly reuiued. Wherein is declared the manner of planting, preserving, and husbanding yong trees of diuers kindes for timber and fuell. And of sowing acornes, chestnuts, beech-mast, the seedes of elmes, ashen-keyes, &c. With the commodities and discommodities of inclosing decayed forrests, commons, and waste grounds. And also the use of a small portable instrument for measuring of board, and the solid content and height of any tree standing. Discoursed in a dialogue betweene a suru-eyour, woodward, gentleman, and a farmer*. London: William Stansby, 1612. STC 4923.

Clark, Stuart. "Inversion, Misrule and the Meaning of Witchcraft." *Past and Present* 87 (May 1980): 98–127.

Cogan, Thomas. *The hauen of health chiefely gathered for the comfort of students, and consequently of all those that haue a care of their health, amplified vpon fiue words of Hippocrates, written Epid. 6 Labor, cibus, potio, somnus, Venus: by Thomas Coghan master of Artes, & Bacheler of Phisicke. Hereunto is added a preseruation from the pestilence, with a short censure of the late sicknes at Oxford.* London: Henrie Midleton, for William Norton, 1584. STC 193:17.

Cohen, Jeffrey Jerome. "Monster Culture (Seven Theses)." *Monster Theory.* Ed. Jeffrey Jerome Cohen. Minneapolis and London: U of Minnesota P, 1996. 3–25.

Cohen, Michael. "Blues in the Green: Ecocriticism under Critique." *Environmental History* 9.1 (January 2004): 9–36.

Cohen, Ted. "Metaphor and the cultivation of intimacy." *On Metaphor.* Ed. Sheldon Sacks. Chicago and London: U of Chicago P, 1978. 1–10.

Compost, E. "What is a weed?" January 13, 2010. Web. January 20, 2010. <http://www.emilycompost.com/weed_definition.htm>

Crosby, Alfred W. *Ecological Imperialism: The Biological Expansion of Europe, 900–1900.* Cambridge: Cambridge UP, 1986.

Culler, Jonathan. *On Deconstruction: Theory and Criticism after Structuralism.* Ithaca, NY: Cornell UP, 1982.

Currie, Dawn H., and Valerie Raoul. "The Anatomy of Gender: Dissecting Sexual Difference in the Body of Knowledge." *The Anatomy of Gender: Women's Struggle for the Body.* Ed. Dawn H. Currie and Valerie Raoul. Ottawa: Carleton UP, 1992. 1–34.

Dam, Julie K. L., Margaret Emery, and Sinting Lai. "Talk of the Streets— Hong Kong: Uncouth Cuisine." *Time* May 1, 1995: 12.

Danby, John. *Shakespeare's Doctrine of Nature: A Study of King Lear.* London: Faber and Faber, 1948.

Dannenfeldt, Karl. H. "Sleep: Theory and Practice in the Late Renaissance." *Journal of the History of Medicine and Allied Sciences* 41 (October 1986): 415–41.

Dekker, Thomas. *Nevves from Graues-end sent to nobody.* London: T[homas] C[reede] for Thomas Archer, 1604. STC 1139:17.

Delaney, Janice, Mary Jane Lupton, and Emily Toth. *The Curse: A Cultural History of Menstruation.* New York: Signet, 1976.

Delany, Paul. "*King Lear* and the Decline of Feudalism." *PMLA* 92.3 (May 1977): 429–40.

de Man, Paul. "The Epistemology of Metaphor." *On Metaphor.* Ed. Sheldon Sacks. Chicago and London: U of Chicago P, 1978. 11–28.

Diamond, Irene, and Gloria Orenstein (eds.). *Reweaving the World: The Emergence of Ecofeminism.* San Francisco: Sierra Club Books, 1990.

Dobson, Mary J. *Contours of Death and Disease in Early Modern England.* Cambridge: Cambridge UP, 1997.

Dollimore, Jonathan. *Radical Tragedy: Religion, Ideology, and Power in the Drama of Shakespeare and His Contemporaries.* 2nd ed. Durham: Duke UP, 1993.

Douglas, Mary. *Purity and Danger: An Analysis of Concepts of Pollution and Taboo*. London: Routledge and Kegan Paul, 1966.

———. "The Idea of Home: A Kind of Space." *Social Research* 58.1 (Spring 1991): 287–307.

Dubrow, Heather. *Captive Victors: Narrative Poems and Sonnets*. Ithaca, NY and London: Cornell UP, 1987.

Du Laurens, Andrea. *A discourse of the preservation of the sight of melancholic diseases of rheumes and of old age*. London: Felix Kingston for Ralph Jackson, 1599. STC 7304.

Egan, Gabriel. *Green Shakespeare: From Ecopolitics to Ecocriticism*. London and New York: Routledge, 2006.

Elliot, Robert. "Introduction: Human-Centred Environmental Ethics." *Environmental Ethics*. Oxford and New York: OUP, 1995. 1–20.

Elton, William R. *King Lear and the Gods*. Lexington: UP of Kentucky, 1988.

Elyot, Thomas, Sir. *The castel of helth corrected and in some places augmented, by the fyrste authour therof, syr Thomas Elyot knyght, the yere of oure lord 1541*. London: Thomae Bertheleti, 1541. STC (2nd ed.) 7644.

Estok, Simon C. "Environmental Implications of the Writing and Policing of the Early Modern Body: Dismemberment and Monstrosity in Shakespearean Drama." *Shakespeare Review* 33 (1998): 107—42.

———. "An Introduction to Shakespeare and Ecocriticism: The Special Cluster." *ISLE: Interdisciplinary Studies in Literature and Environment* 12.2 (Summer 2005): 109–17.

———. "Letter," "Forum on Literatures of the Environment," *PMLA* 114.5 (October 1999): 1095–96.

———. "Reading the 'Other' Where Fancy Is Bred: Designating Strangers in Shakespeare." Ph.D. dissertation, U of Alberta, 1996.

———. "A Report Card on Ecocriticism." *AUMLA: The Journal of the Australasian Universities Language and Literature Association* 96 (November 2001): 220–38.

———. "Theorizing in a Space of Ambivalent Openness: Ecocriticism and Ecophobia." *ISLE*, 16.2 (Spring 2009): 203–25.

———. "Theory from the Fringes: Animals, Ecocriticism, Shakespeare." *Mosaic* 40.1 (March 2007): 61–78.

Evans, K. W. "The Racial Factor in *Othello*." *Shakespeare Studies* 5 (1969): 124–40.

Evelyn, John. *Fumifugium: or, The inconveniencie of the aer and smoak of London dissipated. Together with some Remedies humbly proposed*. (Facsim of 1661 Edition). Exeter: The Rota, 1976.

———. *SILVA, or a Discourse of FOREST-TREES, and the Propagation of Timber in His MAJESTIES Dominions. As it was Deliver'd in the ROYAL SOCIETY the xvth of October, MDCLXII, upon occasion of certain Quæries propounded to that Illustrious Assembly, by the Honourable the Principal Officers, and Commissioners of the Navy. To which is annexed POMONA; Or, An Appendix concerning Fruit-Trees in relation to CIDER; The*

160   WORKS CITED

Making, and severall wayes of Ordering it. Second edition. London: John Martyn and James Allestry, Printers to the Royal Society, 1670. Wing E3517.

Evernden, Neil. *The Social Creation of Nature.* Baltimore: Johns Hopkins UP, 1992.

Fagan, Brian. *The Little Ice Age: How Climate Made History, 1300–1850.* New York: Basic, 2001.

Feder, Helena. " 'Biogenetic Intervention' (Or 'gardening,' Shakespeare, and the Future of Ecological Thought)." *Green Letters: Studies in Ecocriticism* 9 (Spring 2008): 33–47.

Felperin, Howard. " 'Tongue-tied our queen?': The Deconstruction of Presence in *The Winter's Tale.*" *Shakespeare and the Question of Theory.* Ed. Patricia Parker and Geoffrey Hartman. New York: Methuen, 1985. 3–18.

Foucault, Michel. *The History of Sexuality: Volume 1, An Introduction.* Trans. Robert Hurley. New York: Vintage, 1990.

Freed, Eugenie R. "Nature in Shakespeare's King Lear." *Hebrew University Studies in Literature and the Arts* 15 (Fall 1987): 44–54.

Gaard, Greta. "annie potts special issue reading." Email to the author. February 3, 2010.

——— (ed). *Ecofeminism: Women, Animals, Nature.* Philadelphia: Temple UP, 1993.

———. "Toward a Queer Ecofeminism." *Hypatia* 12.1 (Winter 1997): 114–37.

Gaard, Greta, and Patrick Murphy. "Introduction." *Ecofeminist Literary Criticism: Theory, Interpretation, Pedagogy.* Ed. Greta Gaard and Patrick Murphy. Urbana and Chicago: U of Illinois P, 1998. 1–13.

Gallop, Jane. *The Daughter's Seduction: Feminism and Psychoanalysis.* Ithaca, NY: Cornell UP, 1982.

Garrard, Greg. "Green Shakespeares." *British Shakespeare Association Biennial Conference 2005—Seminar: Shakespeare and Ecology.* Newcastle, UK. September 2, 2005.

Garzoni, Tomaso. *Hospitall of Incurable Fooles: Erected in English, as neer the first Italian modell and platforme, as the unskilfull hand of an ignorant Architect could devise.* (Trans. Edward Blount). London: Edm. Bollifant for Edward Blount, 1600. STC 11634.

Gay, F. "The Midland Revolt and the Inquisitions of Depopulation of 1607." *Transactions of the Royal Historical Society, N.S.* 18 (1904): 195–244.

Gibbs, Raymond W., Jr. "Process and products in making sense of tropes." *Metaphor and Thought, 2nd Edition.* Ed. Andrew Ortony. Cambridge: Cambridge UP, 1993. 252–76.

Giffney, Noreen, and Myra J. Hird (eds.). *Queering the Non/Human.* Burlington, VA: Ashgate, 2008.

Gifford, Terry. "Book Review: *Green Shakespeare: From Ecopolitics to Ecocriticism.*" *ISLE: Interdisciplinary Studies in Literature and Environment* 13.2 (Summer 2006): 272–73.

Goldberg, Jonathan. "The Anus in *Coriolanus*." *Historicism, Psychoanalysis, and Early Modern Culture*. Ed. Carla Mazzio and Douglas Trevor. New York: Routledge, 2000. 260–71.

———. *Sodometries: Renaissance Texts, Modern Sexualities*. Stanford, CA: Stanford UP, 1992.

Goss, Robert. *Jesus Acted Up: A Gay and Lesbian Manifesto*. New York: HarperSanFrancisco, 1994.

Gouge, William. *Dearth's Death: Or, A Removall of Famine: Gathered out of II Sam. XXI.1*. London: George Miller for Edward Brewster, 1631. STC 12116.

Grady, Hugh, "*Hamlet* and the present," *Presentist Shakespeares*. Ed. Hugh Grady and Terrence Hawkes. New York, Routledge, 2007. 141–63.

———. "Why Presentism Now," *SHAKSPER: The Global Electronic Shakespeare Conference*, January 29, 2007. Web. June 1, 2009. <http://www.shaksper.net/archives/2007/0065.html>

Graunt, John. *Natural and Political Observations Mentioned in a following INDEX, and made upon the Bills of Mortality*. London: John Martyn, Printer to the Royal Society, 1676. Wing G1602.

Greenblatt, Stephen. "Filthy Rites." *Learning to Curse: Essays in Early Modern Culture*. New York and London: Routledge, 1992. 59–79.

———. "Murdering peasants: status, genre, and the representation of rebellion." *Learning to Curse: Essays in Early Modern Culture*. New York and London: Routledge, 1992. 99–130.

———. *Renaissance Self-Fashioning: From More to Shakespeare*. Chicago and London: U of Chicago P, 1980.

———. "Shakespeare and the Exorcists." *Shakespearean Negotiations: The Circulation of Social Energy in Renaissance England*. Berkeley and Los Angeles: U of California P, 1988. 94–128.

———. *Will in the World: How Shakespeare Became Shakespeare*. New York: Norton, 2005.

Gregory, Derek. *Geographical Imaginations*. Cambridge, MA and Oxford, UK: Blackwell, 1994.

Griffin, Susan. *Woman and Nature: The Roaring Inside Her*. New York: Harper and Row, 1978.

Hale, David George. *The Body Politic: A Political Metaphor in Renaissance Literature*. The Hague and Paris: Mouton, 1971.

Haraway, Donna J. *Simians, Cyborgs, and Women: The Reinvention of Nature*. New York: Routledge, 1991.

Hart, James. *Klinike, or the Diet of the Diseased. Divided into Three Books. Wherein is set downe at length the whole matter and nature of Diet for those in health, but especially for the sicke; the Aire, and other Elements: Meat and Drinke, with divers other things; various controversies concerning this subject are discussed*. London: John Beale for Robert Allot, 1633. STC 12888.

Hawkes, Terence. "Swisser-Swatter: making a man of English letters." *Alternative Shakespeares*. Ed. John Drakakis. London and New York: Routledge, 1985. 26–46.

Heise, Ursula. *Sense of Place, Sense of Planet: The Environmental Imagination of the Global*. New York: Oxford UP, 2008.

Hendricks, Margo. "Surveying 'Race' in Shakespeare." *Shakespeare and* Race. Ed. Catherine M. S. Alexander and Stanley Wells. New York: Cambridge UP, 2000. 1–22.

Heywood, Thomas. *The Fair Maid of the West, Parts I & II*. Ed. Robert K. Turner. Lincoln: U of Nebraska P, 1967.

Hillard, Tom J. "'Deep into that darkness peering': An Essay on Gothic Nature." *ISLE: Interdisciplinary Studies in Literature and Environment* 16.4 (Autumn 2009): 685–95.

Hiltner, Ken. "Renaissance Literature and Our Contemporary Attitude toward Global Warming." *ISLE: Interdisciplinary Studies in Literature and Environment* 16.3 (Summer 2009): 429–41.

Hodges, Devon L. *Renaissance Fictions of Anatomy*. Amherst: U of Massachusetts P, 1985.

Honour, Hugh. *The New Golden Land: European Images of America from the Discoveries to the Present Time*. New York: Pantheon, 1975.

Horace. *The Odes and Epodes of Horace*. Ed. and trans. Joseph P. Clancy. Chicago and London: U of Chicago P, 1960.

Horrox, Rosemary (Ed. and Trans). *The Black Death*. Manchester and New York: Manchester UP, 1994.

Huggan, Graham, and Helen Tiffin. *Postcolonial Ecocriticism: Literature, Animals, Environment*. London and New York: Routledge, 2010.

Hulme, Peter. *Colonial Encounters: Europe and the Native Caribbean, 1492–1797*. London and New York: Methuen, 1986.

———. "Columbus and the Cannibals: A Study of the Reports of Anthropophagy in the Journal of Christopher Columbus." *Ibero-Amerikanisches Archiv* 4 (1978): 115–39.

Hunt, Maurice. "The Backward Voice of Coriol-anus." *Shakespeare Studies* 32 (2004): 220–39.

Jagendorf, Zvi. "Coriolanus: Body Politic and Private Parts." *Shakespeare Quarterly* 41.4 (Winter 1990): 455–69.

James, N.D.G. *A History of English Forestry*. Oxford: Blackwell, 1981.

Jones, Eldred. *Othello's Countrymen: The African in English Renaissance Drama*. London: Oxford UP, 1965.

Jordan, Constance. "'Eating the Mother': Property and Propriety in *Pericles*." *Creative Imitation: New Essays on Renaissance Literature in Honor of Thomas M. Greene*. Ed. David Quint, Margaret W. Ferguson, G. W. Pigman III, and Wayne A. Rebhorn. Binghamton, NY: Medieval and Renaissance Texts and Studies, Vol. 95, 1992. 333–53.

Jorden, Edward. *A briefe discourse of a disease called the suffocation of the mother... Wherin is declared that diuers strange actions and passions of the body of man, which in the common opinion, are imputed to the diuell, haue their true naturall causes*. London: Iohn Windet, 1603. STC 14790.

Kahn, Coppélia. *Man's Estate: Masculine Identity in Shakespeare*. Berkeley: U of California P, 1981.

Kamps, Ivo, Thomas Hallock, and Karen Raber (eds.). *Early Modern Ecostudies: From the Florentine Codex to Shakespeare.* New York: Palgrave Macmillan, 2008.

Kehler, Dorothea. "Teaching the Slandered Women of *Cymbeline* and *The Winter's Tale.*" *Approaches to Teaching Shakespeare's* The Tempest *and Other Late Romances.* Ed. Maurice Hunt. New York: Modern Language Association, 1992. 80–86.

Kerridge, Richard. "Introduction." *Writing the Environment: Ecocriticism and Literature.* Ed. Richard Kerridge and Neil Sammells. London and New York: Zed Books, 1998. 1–9.

Kiefer, Frederick. "Art, Nature, and the Written Word in *Pericles.*" *University of Toronto Quarterly: A Canadian Journal of the Humanities* 61.2 (Winter 1991–92): 207–25.

Kilgour, Maggie. *From Communion to Cannibalism: An Anatomy of Metaphors of Incorporation.* Princeton, NJ: Princeton UP, 1990.

King, Ynestra. "Healing the Wounds: Feminism, Ecology, and the Nature/ Culture Dualism." *Reweaving the World: The Emergence of Ecofeminism.* Ed. Irene Diamond and Gloria Feman Orenstein. San Francisco: Sierra Club, 1990. 106–21.

———. "Toward an Ecological Feminism and a Feminist Ecology." *Machina Ex Dea: Feminist Perspectives on Technology.* Ed. Joan Rothschild. New York: Pergamon, 1983. 118–129.

Kleinberg, Seymour. "The Merchant of Venice: The Homosexual as Anti-Semite in Nascent Capitalism." *Journal of Homosexuality* 8 (1983): 113–26.

Kolodny, Annette. *The Lay of the Land: Metaphor as Experience and History in American Life and Letters.* Chapel Hill: U of North Carolina P, 1975.

Kristeva, Julia. *Powers of Horror: An Essay on Abjection.* Trans. Leon S. Roudiez. New York: Columbia UP, 1982.

Laird, David. "Competing Discourses in The Winter's Tale." *Connotations* 4.1–2 (1994–95): 25–43.

Lakoff, George, and Mark Johnson. *Metaphors We Live By.* Chicago and London: U of Chicago P, 1980.

Laroque, François. "The Jack Cade Scenes Reconsidered: Popular Rebellion, Utopia, or Carnival?" *Shakespeare and Cultural Traditions.* Ed. Tetsuo Kishi, Roger Pringle, and Stanley Wells. Newark and London: U of Delaware P and Associated UP, 1994. 76–89.

Laslett, Peter. *The World We Have Lost.* London: Methuen, 1965.

Latour, Bruno. *We Have Never Been Modern.* Trans. Catherine Porter. Cambridge, MA: Harvard UP, 1993.

Lee, Wendy Lynne, and Laura M. Dow, "Queering Ecological Feminism: Erotophobia, Commodification, Art, and Lesbian Identity," *Ethics and the Environment* 6.2 (Autumn 2001): 1–21.

Leggatt, Alexander. "The Shadow of Antioch: Sexuality in Pericles and Prince of Tyre." *Parallel Lives: Spanish and English National Drama, 1580–1680.*

Ed. Louise Fothergill-Payne and Peter Fothergill-Payne. Cranbury, NJ. Associated UP, 1991. 167–79.

Leiss, William. *The Domination of Nature.* Montréal: McGill/Queens UP, 1994.

Leiss, William, Stephen Kline, and Sut Jhally. *Social Communication in Advertising: Persons, Products, and Images of Well-Being.* New York and London: Methuen, 1986.

Levin, Richard. "Caught Napping or Sleeping on the Job in Shakespeare." *Shakespeare Newsletter* 57: 1. No. 271 (Spring/Summer 2007): 13–14.

Levin, Samuel R. "Language, Concepts, and Works: Three Domains of Metaphor." *Metaphor and Thought, 2nd Edition.* Ed. Andrew Ortony. Cambridge: Cambridge UP, 1993. 112–23.

Levy, Neil. "Foucault's Unnatural Ecology." *Discourses of the Environment.* Ed. Eric Darier. Oxford. UK: Blackwell, 1999. 202—16.

Lewin, Jennifer. "'Your Actions Are My Dreams': Sleepy Minds in Shakespeare's Last Plays." *Shakespeare Studies* 31 (2003): 184–204.

Lewis, Anthony J. "'I Feed on Mother's Flesh': Incest and Eating in *Pericles.*" *Essays in Literature* 15.2 (1988): 147–63.

Little, Arthur L. Jr. "'Transshaped' Women: Virginity and Hysteria in *The Changeling.*" *Madness in Drama—Themes in Drama* 15. Cambridge: Cambridge UP, 1993. 19–42.

Lodge, Thomas. *A Treatise of the Plague: Containing the nature, signes, and accidents of the same, with the certaine and absolute cure of the Feuers, Botches and Carbuncles that raigne in these times: And aboue all things most singular Experiments and preseruatiues in the same, gathered by the obseruation of diuers worthy Trauailers, and selected out of the writings of the best learned Phisitians in this age.* London: Edward White and N.L., 1603. STC 16676.

Loomba, Ania. "The Color of Patriarchy: Critical Difference, Cultural Difference, and Renaissance Drama." *Women, "Race," and Writing in the Early Modern Period.* Ed. Margo Hendricks and Patricia Parker. London and New York: Routledge, 1994. 17–34.

Lowe, Lisa. "'Say I play the man I am': Gender and Politics in Coriolanus." *Kenyon Review* 8.4 (Fall 1986): 86–95.

Lykosthenes, Konrad. *The doome warning all men to the iudgemente: wherein are contayned for the most parte all the straunge prodigies hapned in the worlde, with diuers secrete figures of reuelations tending to mannes stayed conuersion towards God: in maner of a generall chronicle, gathered out of sundrie approued authors by St. Batman professor in diuinite [Prodigiorum ac ostentorum chronicon].* London, 1557. GR825.L8 1557 Cage Folio.

Maguin, Jean-Marie. "Rise and Fall of the King of Darkness." *French Essays on Shakespeare and His Contemporaries: "What would France with us?"* Ed. Jean-Marie Maguin and Michele Willems. Newark: U of Delaware P; London: Associated UP, 1995. 247–70.

Manes, Christopher. "Nature and Silence." *The Ecocriticism Reader: Landmarks in Literary Ecology.* Ed. Cheryll Glotfelty and Harold Fromm. Athens: U of Georgia P, 1996: 15—29.

Manwood, John. *A Treatise and Discourse of the Lawes of the Forrest: Wherein is declared not onely those lawes, as they are now in force, but also the originall and beginning of Forrestes: And what a forrest is in his owne proper nature, and wherein the same doth differ from a Chase, a Park, or a Warren, with all such thinges as are incident or belonging thereunto, at large doth appeare in the Table in the beginning of this Booke.* London: Thomas Wight and Bonham Norton, 1598. STC 17291.

Markley, Robert. "Summer's Lease: Shakespeare in the Little Ice Age." *Early Modern Ecostudies: From the Florentine Codex to Shakespeare.* Ed. Ivo Kamps, Thomas Hallock, and Karen Raber. New York: Palgrave Macmillan, 2008. 131–42.

Marx, Karl. "Part 1: Commodities and Money." *Capital: A Critique of Political Economy, Vol. 1.* Trans. Samuel Moore and Edward Aveling. Ed. Frederick Engels. Moscow: Progress Publishers, 1954. 43–144.

Mazel, David. "Ecocriticism as Praxis." *Teaching North American Environmental Literature.* Ed. Laird Christensen, Mark C. Long, and Fred Waage. New York: MLA, 2008. 37–43.

McLuskie, Kathleen. "'The Future in an Instant.'" *Presentism, Gender, and Sexuality in Shakespeare.* Ed. Evelyn Gajowski. New York, Palgrave Macmillan, 2006. 239–51.

Meeker, Joseph. *The Comedy of Survival: Literary Ecology and a Play Ethic, 3rd edition.* Tucson: U of Arizona P, 1997.

Mellor, Mary. *Feminism and Ecology.* Washington Square, New York: New York UP, 1997.

Merchant, Carolyn. *The Death of Nature: Women, Ecology and the Scientific Revolution.* New York: HarperCollins, 1990.

Middleton, Thomas, and William Rowley. *The Changeling.* Ed. George Walton Williams. Lincoln and London: U of Nebraska P, 1966.

Mies, Maria. "White Man's Dilemma: His Search for What He Has Destroyed." *Ecofeminism.* Ed. Maria Mies and Vandana Shiva. London and New Jersey: Zed Books, 1993. 132–63.

Miller, William Ian. *The Anatomy of Disgust.* Cambridge, MA: Harvard UP, 1997.

Montaigne, M. E. "On Cannibals." *Montaigne: Essays.* New York: Penguin, 1958. 105–19.

Montrose, Louis. *The Purpose of Playing: Shakespeare and the Cultural Politics of the Elizabethan Theatre.* Chicago and London: U of Chicago P, 1996.

Moore, Peter. "The Nature of King Lear." *English Studies* 87.2 (April 2006): 169–190.

Moore, Richard B. "Carib 'Cannibalism': A Study in Anthropological Stereotyping." *Caribbean Studies* 13 (1973): 117–35.

Morton, Timothy. *Ecology without Nature: Rethinking Environmental Aesthetics*. Cambridge, MA and London, England: Harvard UP, 2007.

Moulton, Thomas. *This is the myrour or glasse of helthe necessary and nedefull for euery persone to loke in, that wyll kepe theyr body frome the syckenes of the pestile[n]ce? And it sheweth howe the planettes reygne in euery houre of the daye and nyght, with the natures and exposicio[n] of the. xij. signes, deuyded by the. xij. monthes of the yere, [and] sheweth the remedyes for many diuers infirmites [and] dyseases that hurteth the body of man*. London: Robert Redman, 1540. STC (2nd ed.) 18216.

Mullaney, Steven. *The Place of the Stage: License, Play, and Power in Renaissance England*. Chicago and London: U of Chicago P, 1988.

Münster, Sebastian. *Cosmographia [Cosmographiae uniuersalis lib. VI.: in quibus, iuxta certioris fidei scriptorum traditionem describuntur:omniu[m] habitabilis orbis partiu[m] situs, propriae[que] dotes, regionum topographicae effigies...item omnium gentium mores, leges, religio...atque memorabilium in hunc usque annum 1554 gestarum rerum historia]*. Basileae: Henrichum Petri, September 1554. G113 .M7 1554 Cage.

Neely, Carol Thomas. *Distracted Subjects: Madness and Gender in Shakespeare and Early Modern Culture*. Ithaca, NY: Cornell UP, 2004.

———. " 'Documents in Madness': Reading Madness and Gender in Shakespeare's Tragedies and Early Modern Culture." *Shakespeare Quarterly* 42.3 (1991): 315–38.

Nevo, Ruth. "The Perils of *Pericles*." *The Undiscover'd Country: New Essays on Psychoanalysis and Shakespeare*. Ed. B. J. Sokol. London: Free Association P, 1993. 150–178.

Newman, Karen. *Fashioning Femininity and English Renaissance Drama*. Chicago and London: U of Chicago P, 1991.

———. " 'And Wash the Ethiop White': Femininity and the Monstrous in Othello." *Shakespeare Reproduced: The Text in History and Ideology*. Ed. Jean E. Howard and Marion F. O'Connor. New York and London: Routledge, 1987. 142–62.

Nixon, Anthony. *The Dignitie of Man, Both in the perfections of his soule and bodie. Shewing as well the faculties in the disposition of the one: as the senses and organs, in the composition of the other*. London: Edward Allde, 1612. STC 18584.

O'Dair, Sharon. *Class, Critics, and Shakespeare: Bottom Lines on the Culture Wars*. Ann Arbor: U of Michigan P, 2000.

———. "Is It Shakespearean Ecocriticism If It Isn't Presentist?" *Ecocritical Shakespeare*. Ed. Lynne Bruckner and Dan Brayton. Aldershot, England: Ashgate, 2011. 71–85.

———. "Notes." Email to author. November 10, 2008.

———. "Slow Shakespeare: An Eco-critique of 'Method' in Early Modern Literary Studies." *Early Modern Ecostudies: From the Florentine Codex to Shakespeare*. Ed. Ivo Kamps, Karen Raber, and Thomas Hallock. New York: Palgrave Macmillan, 2008. 11–30.

———. "The State of the Green: A Review Essay on Shakespearean Ecocriticism." *Shakespeare* (1745–0926) 4.4 (December 2008): 475–93.

Orgel, Stephen. "Nobody's Perfect: Or Why Did the English Stage Take Boys for Women?" *South Atlantic Quarterly* 88.1 (Winter 1989): 7–29.

———. "Shakespeare and the Cannibals." *Cannibals, Witches, and Divorce: Estranging the Renaissance.* Ed. Marjorie Garber. Baltimore and London: Johns Hopkins UP, 1985. 40–66.

Ortony, Andrew. "Metaphor, Language, and Thought." *Metaphor and Thought, 2nd Edition.* Ed. Andrew Ortony. Cambridge: Cambridge UP, 1993. 1–16.

Oster, Emily. "Witchcraft, Weather and Economic Growth in Renaissance Europe." *Journal of Economic Perspectives* 18.1 (Winter 2004): 215–28.

Paré, Ambroise. *On Monsters and Marvels.* Trans. with introduction and notes by Janis L. Pallister. Chicago and London: U of Chicago P, 1982.

Parham, John. "The Poverty of Ecocritical Theory: E. P. Thompson and the British Perspective." *New Formations: A Journal of Culture/Theory/Politics—Special Issue: Earthographies: Ecocriticism and Culture* 64 (Spring 2008): 25–36.

Park, Katharine, and Lorraine J. Daston. "Unnatural Conceptions: The Study of Monsters in Sixteenth- and Seventeenth-Century France and England." *Past and Present: A Journal of Historical Studies* 92 (August 1981): 21–54.

Parker, Patricia. "Fantasies of 'Race' and 'Gender': Africa, Othello and Bringing to Light." *Women, "Race," and Writing in the Early Modern Period.* Ed. Margo Hendricks and Patricia Parker. London and New York: Routledge, 1994. 84–100.

———. *Literary Fat Ladies: Rhetoric, Gender, Property.* London and New York: Methuen, 1987.

Paster, Gail Kern. *The Body Embarrassed: Drama and the Disciplines of Shame in Early Modern England.* Ithaca, NY: Cornell UP, 1993.

Patterson, Annabel. *Shakespeare and the Popular Voice.* Oxford: Basil Blackwell, 1989.

Peers, Edgar Allison. *Elizabethan Drama and Its Mad Folk.* Cambridge: W. Heffer and Sons, 1914.

Pettet, E. C. "*Coriolanus* and the Midlands Insurrection of 1607." *Shakespeare Survey* 3 (1950): 34–42.

Phillips, Dana. *The Truth of Ecology: Nature, Culture, and Literature in America.* Oxford, Oxford UP, 2003.

Polk, Danne. "Deconstructing Origins: Preliminaries for a Queer Ecological Identity Theory," *5th Annual Symposium on Lesbian, Gay, Bisexual and Transgender Issues,* U of Rhode Island, March, 1999. Web. January 6, 2008. <http://www.queertheory.com/theories/science/deconstructing_origins.htm>

Pringle, James. "Famine Drives North Koreans to Cannibalism." *London Times.* April 13, 1998. Web. January 15, 2001. <http://www.the-times.

co.uk/news/pages/tim/98/04/13/timfgnfar02001.html?1124027>
Link broken

Pugliatti, Paola. " 'More than History Can Pattern': The Jack Cade Rebellion in Shakespeare's *Henry VI, 2*." *Journal of Medieval and Renaissance Studies* 22.3 (Fall 1992): 451–78.

Raber, Karen. "Recent Ecocritical Studies of English Renaissance Literature." *ELR: English Literary Renaissance* 37.1 (February 2007): 151–71.

Read, Daphne. "Introduction to 'Master Harold'...and the Boys." *The Harbrace Anthology of Literature*. Ed. Jon Stott, Raymond Jones, and Rick Bowers. Toronto: Harcourt Brace, 1993. 1257–60.

Regina v. Litchfield, 4—86 *Canadian Criminal Code (C.C.C.)*. (3d). November 18, 1993. Web. October 15, 2009. <http://csc.lexum.umontreal.ca/en/1993/1993scr4–333/1993scr4–333.html>

Richards, I. A. "Metaphor." *The Philosophy of Rhetoric*. London: Oxford UP, 1936.

Riss, Arthur. "The Belly Politic: Coriolanus and the Revolt of Language." *ELH* 59.1 (Spring 1992): 53–75.

Roberts, David. "Sleeping Beauties: Shakespeare, Sleep and the Stage." *Cambridge Quarterly* 35.3 (2006): 231–54.

Roberts, Jeanne Addison. *The Shakespearean Wild: Geography, Genus, and Gender*. Lincoln and London: U of Nebraska P, 1991.

Robertson, Jean. "*Macbeth* on Sleep: 'Sore Labour's Bath' and Sidney's 'Astrophil and Stella,' XXXIX." *Notes and Queries* 14 (1967): 139–41.

Rose, Gillian. *Feminism and Geography: The Limits of Geographical Knowledge*. Minneapolis: U of Minnesota P, 1993.

Roszak, Theodore. *The Voice of the Earth: An Exploration of Ecopsychology*. Grand Rapids, MI: Phanes P, 2001.

Rueff, Jacob (Rüff, Jakob). *The expert midwife, or An excellent and most necessary treatise of the generation and birth of man Wherein is contained many very notable and necessary particulars requisite to be knovvne and practised: with diuers apt and usefull figures appropriated to this worke. Also the causes, signes, and various cures, of the most principall maladies and infirmities incident to women. Six bookes compiled in Latine by the industry of Iames Rueff, a learned and expert chirurgion: and now translated into English for the generall good and benefit of this nation*. London: E. G[riffin], 1637. STC (2nd ed.) 21442.

Rumelhart, David E. "Some Problems with the Notion of Literal Meanings." *Metaphor and Thought, 2nd Edition*. Ed. Andrew Ortony. Cambridge: Cambridge UP, 1993. 71–82.

Ryan, Kiernan. "*Troilus and Cressida*: The Perils of Presentism." *Presentist Shakespeares*. Ed. Hugh Grady and Terrence Hawkes. New York, Routledge, 2007. 164–83.

Salkeld, Duncan. *Madness and Drama in the Age of Shakespeare*. Manchester and New York: Manchester UP, 1993.

Sandilands, Catriona. "From Unnatural Passions to Queer Nature." *Alternatives Journal* 27.3 (2001): 31–35.

Scarry, Elaine. *The Body in Pain: The Making and Unmaking of the World.* New York and Oxford: Oxford UP, 1985.

Schedel, Hartmann. *Liber chronicarum [Registrum huius operis Libri cronicarum cum figuris et ijmagibus ab inicio mundi].* Anthonius Koberger: Nuremberg, July 1493. INC S281.

Schleiner, Winfried. "Burton's Use of praeteritio in Discussing Same-Sex Relationships." *Renaissance Discourses of Desire.* Ed. Claude J. Summers and Ted-Larry Pebworth. Columbia and London: U of Missouri P, 1993. 159–78.

Searle, John R. "Metaphor." *Metaphor and Thought, 2nd Edition.* Ed. Andrew Ortony. Cambridge: Cambridge UP, 1993. 83–111.

Shakespeare, William. *The Riverside Shakespeare, 2nd Edition.* Ed. G. Blakemore Evans and J. J. M. Tobin. Boston and New York: Houghton Mifflin, 1997.

Shanker, Sidney. "Some Clues for Coriolanus." *Shakespeare Association Bulletin* 24 (1949): 209–13.

Sillars, Les. "'Gay,' Perhaps, But Far from Happy." *Alberta Report* 22.26 (June 12, 1995): 30–31.

Sinsheimer, Hermann. *Shylock, the History of a Character.* New York: Benjamin Blom, 1963.

Slemon, Stephen. "Bones of Contention: Post-Colonial Writing and the 'Cannibal' Question." *Literature and the Body.* Ed. Anthony Purdy. Amsterdam and Atlanta, GA: Rodopi, 1992: 163–77.

Slovic, Scott. "Foreword." *The Greening of Literary Scholarship* Iowa City: U of Iowa Press, 2002: vii–xi.

———. "Further reflections." Email to the author. December 6, 2009.

———. "Re: LIKELY SPAM I'm Not Sure If You Are Receiving." Email to the author. September 16, 2008.

Smith, Hallett. "Pericles, Prince of Tyre." *The Riverside Shakespeare, 2nd Edition.* Ed. G. Blakemore Evans and J. J. M. Tobin. Boston and New York: Houghton Mifflin, 1997. 1527–30.

Smith, Mick. *An Ethics of Place.* Albany, NY: SUNY, 2001.

Sobel, David. *Beyond Ecophobia: Reclaiming the Heart in Nature Education.* Great Barrington, MA: Orion, 1996.

Sparke, Michael, Jr. *The Drousie Disease; or, An Alarme to Awake Churchsleepers.* London: I.D. for Michael Sparke, 1638. STC 254.8.

Stallybrass, Peter. "*Macbeth* and Witchcraft. *Focus on* Macbeth." Ed. John Russell Brown. London, Boston, and Henley: Routledge and Kegan Paul, 1982. 189–209.

———. "Patriarchal Territories: The Body Enclosed." *Rewriting the Renaissance: The Discourses of Sexual Difference in Early Modern Europe.* Ed. Margaret W. Ferguson, Maureen Quilligan, and Nancy J. Vickers. Chicago and London: U of Chicago P, 1986. 123–42.

———. "Shakespeare, the Individual, and the Text." *Cultural Studies.* Ed. Lawrence Grossberg, Cary Nelson, and Paula Treichler. London and New York: Routledge, 1992. 593–612.

Standish, Arthur. *The Commons Complaint. Wherein is contained two speciall grievances. The first, the general destruction and waste of woods in this Kingdome, with a remedie for the same: Also how to plant wood according to the nature of euery soile, without losse of ground, and how thereby many more, and better Cattell may be yearely bred, with the charge and profit that yearely may arise thereby. The second grievance is, the extreame dearth of Victualls.* London: William Stansby. 1612. STC 23203.

St. John, Spencer Sir. "Philosophy: Character—Physical Science—Sleep and Dreams." *Essays on Shakespeare and His works.* London: Smith, Elder, 1908: 182–98.

Stockholder, Kay. "Sex and Authority in *Hamlet, King Lear,* and *Pericles.*" *Mosaic* 18.3 (Fall 1985): 17–29.

Stolz, Kit. "Ecophobia: A Paradigm." *A Change in the Wind* November 25, 2005. March 1, 2008. <http://achangeinthewind.typepad.com/achangeinthewind/2005/11/ecophobia_a_par.html>

Strickler, Breyan. "Sex and the City: An Ecocritical Perspective on the Place of Gender and Race in *Othello.*" *ISLE: Interdisciplinary Studies in Literature and Environment* 12.2 (Summer 2005): 119–37.

Strong, Roy C. *Portraits of Queen Elizabeth I.* Oxford: OUP, 1963.

Sturgeon, Noël. *Ecofeminist Natures: Race, Gender, Feminist Theory and Political Action.* New York and London: Routledge, 1997.

Tennenhouse, Leonard. *Power on Display: The Politics of Shakespeare's Genres.* New York and London: Methuen, 1986.

Thomas, Keith. *Man and the Natural World: Changing Attitudes in England, 1500–1800.* London: Allen Lane, 1983.

Thompson, Ann, and John Thompson. *Shakespeare: Meaning and Metaphor.* Sussex, England: Harvester, 1987.

Tiffin, Chris, and Alan Lawson. "Introduction: The Textuality of Empire." *De-Scribing Empire: Post-Colonialism and Textuality.* Ed. Chris Tiffin and Alan Lawson. New York and London: Routledge, 1994. 1–11.

Tilley, Christopher. *Metaphor and Material Culture.* Oxford, UK: Blackwell, 1999.

Tilney, Edmund. *A Brief and Pleasant Discourse of Duties in Mariage, called the Flower of Friendshippe.* London: Henrie Denham, 1568. STC 24076.

Tryon, Thomas. *The Way to Health, Long Life and Happiness: Or, A Discourse of Temperance, and the Particular Nature of all Things Requisite for the Life of Man; As, All Sorts of Meats, Drinks, Air, Exercise, &c. with special Directions how to use each of them to the best Advantage of the Body and Mind, Shewing the true ground of Nature, whence most Diseases proceed, and how to prevent them .* London: R. Norton, 1697. Wing 3202.

Tuan, Yi-Fu. *Topophilia: A Study of Environmental Perception, Attitudes, and Values.* New York: Columbia UP, 1974.

Turner, William. *A new herball wherein are conteyned the names of herbes in Greke, Latin, Englysh, Duch [sic] Frenche, and in the potecaries and herbaries Latin, with the properties degrees and naturall places of the same, gathered and made by Wylliam Turner, physicion vnto the Duke of Somersettes Grace.* London: Steven Mierdman, 1551. STC (2nd ed.) 24365.

Twine, Robert. "Intersectional Disgust?—Animals and (Eco) Feminism." *Feminism and Psychology* 20 (2010): 397–406.

van der Straet, Jan. *Nova Reperta.* Plate 1: *America.* Antwerp: Engraved by Theodore and Phillippe Galle, c. 1600. Art vol. f81, plate 1.

van Tine, Robin. "Gaeaphobia: Ecophobia, Ecomania and 'Otherness' in the Late 20th Century." *From Method to Madness: Five Years of Qualitative Enquiry.* U of the Witwatersrand: History of the Present Press, Department of Psychology, 1999. Posted on *Gatherings: Seeking Ecopsychology* Spring issue, May 2000. Web. March 3, 2008. <http://www.ecopsychology.org/journal/gatherings2/robin.htm>

Venner, Tobias. *Via recta ad vitam longam. Or, A plain philosophicall demonstration of the nature, faculties, and effects of allsuch things as by way of nourishments make for the preservation of health, with divers necessary dieteticall observations; as also of the true use and effects of sleep, exercise, excretions, and perturbations, with just applications to every age, constitution of body, and time of yeere.* London: R. Bishop, for Henry Hood, 1637. STC 24646.

Vickers, Nancy. " 'The blazon of sweet beauty's best': Shakespeare's Lucrece." *Shakespeare and the Question of Theory.* Ed. Patricia Parker and Geoffrey Hartman. New York and London: Methuen, 1985. 95–115.

———. "Diana Described: Scattered Woman and Scattered Rhyme." *Critical Inquiry* 8 (1981): 265–79.

———. "This Heraldry in Lucrece' Face." *The Female Body in Western Culture: Contemporary Perspectives.* Ed. Susan Rubin Suleiman. Cambridge, MA and London: Harvard UP, 1986: 209–22.

Vieth, Ilza. *Hysteria: The History of a Disease.* Chicago: U of Chicago P, 1965.

Waage, Frederick O. "Shakespeare Unearth'd." *ISLE: Interdisciplinary Studies in Literature and Environment* 12.2 (Summer 2005): 139–64.

Wadiwel, Dinesh. "Animal by Any Other Name? Patterson and Agamben Discuss Animal (and Human) Life." *Borderlands* 3.1 (2004): 1–34. Web January 6, 2008. <http://www.borderlandsejournal.adelaide.edu.au/vol3no1_2004/wadiwel_animal.htm>

Warren, Karen J. "Introduction." *Ecofeminism: Women, Culture, Nature.* Ed. Karen J. Warren and Nisvan Erkal. Bloomington and Indianapolis: Indiana UP, 1997. xi–xvi.

Watson, Robert N. *Back to Nature: The Green and the Real in the Late Renaissance.* Philadelphia: U of Pennsylvania P, 2006.

Wear, Andrew. "Explorations in Renaissance Writings on the Practice of Medicine." *The Medical Renaissance of the Sixteenth Century.* Ed. A. Wear, R. K. French, and A. M. Lonie. Cambridge: Cambridge UP, 1985. 118–45.

———. "Making Sense of Health and the Environment in Early Modern England." *Medicine in Society: Historical Essays.* Ed Andrew Wear. Cambridge; New York: Cambridge UP, 1992. 119–47.

Webster, John. *The White Devil.* Ed. Christina Luckyj. London and New York: A & C Black and Norton, 1996.

Wells, Robin Headlam. "'Manhood and Chevalrie': *Coriolanus*, Prince Henry, and the Chivalric Revival." *Review of English Studies: A Quarterly Journal of English Literature and the English Language* 51.203 (August 2000): 395–422.

Whitney, Charles. "Dekker's and Middleton's Plague Pamphlets as Environmental Literature." *Representing the Plague in Early Modern England.* Ed. Rebecca Totaro and Ernest Gilman. New York: Routledge, 2010. 201—18.

Williams, George Walton. "Shakespeare's Metaphors of Health: Food, Sport, and Life-Preserving Rest." *Journal of Medieval and Renaissance Studies* 14.2 (Fall 1984): 187–202.

———. "Sleep in *Hamlet*." *Renaissance Papers 1964.* Ed. S. K. Heninger, Jr., Peter G. Phialas, and George Walton Williams. Durham, NC: Southeastern University Conference, 1965. 17–20.

Williams, Raymond. *The Country and the City.* New York: Oxford UP, 1973.

———. *Keywords.* London: Fontana, 1976.

Wilson, Edward O. "Biophilia and the Conservation Ethic." *The Biophilia Hypothesis.* Ed. Stephen R. Kellert and Edward O. Wilson. Washington, DC: Island Press, 1993. 32–41.

Wilson, Robert Rawdon. *The Hydra's Tale Imagining Disgust.* Edmonton: U of Alberta P, 2002.

Wittkower, Rudolf. "Marvels of the East: A Study in the History of Monsters." *Journal of the Warburg and Courtauld Institute* 5 (1942): 159–97.

Wolfe, Cary. *Animal Rites: American Culture, the Discourse of Species, and Posthumanist Theory.* Chicago: U of Chicago P, 2003.

Woodard, Joe. "Victims at Last." *Alberta Report* 22.26 (June 12, 1995): 28–33.

Woodbridge, Linda. *The Scythe of Saturn: Shakespeare and Magical Thinking.* Urbana and Chicago: U of Illinois P, 1994.

———. *Vagrancy, Homelessness, and English Renaissance Literature.* Urbana and Chicago: U of Illinois P, 2001.

———. *Women and the English Renaissance: Literature and the Nature of Womankind, 1540–1620.* Urbana and Chicago: U of Illinois P, 1984.

Wynne-Davies, Marion. "'The Swallowing Womb': Consumed and Consuming Women in *Titus Andronicus*." *The Matter of Difference: Materialist Feminist Criticism of Shakespeare.* Ed. Valerie Wayne. Ithaca, NY: Cornell UP, 1991. 129–51.

Yarranton, Andrew. *England's Improvement by Sea and Land. To Out-do the Dutch without Fighting, to Pay Debts without Moneys, To set at Work all the POOR of England with the Growth of our own Lands. To prevent unnecessary SUITS in Law; With the Benefit of a Voluntary REGISTER. Directions where vast quantities of Timber are to be had for the Building of SHIPS; With the Advantage of making the Great RIVERS of England Navigable. RULES to prevent FIRES in London, and other Great CITIES; With Directions how the several Companies of Handicraftsmen in London may always have cheap Bread and Drink.* London: R. Everingham, 1677. Wing Y13.

Youngs, Frederic A. *The Proclamations of the Tudor Queens.* Cambridge: Cambridge UP, 1976.

Ziegler, Georgianna. "My Lady's Chamber: Female Space, Female Chastity in Shakespeare." *Textual Practice* 4.1 (Spring 1990): 73–90.

INDEX

Bolded page numbers refer to illustrations.

identity
  environmental embeddedness
    and, 13
  land and, 23
  masculine, 21–23, 54, 107, 138n5
  sexual, 35
illness. *see* disease
incest, 15, 79, 81
individualism, 21, 36, 44, 62,
    113–14, 135n4
Industrial Revolution, 7
inheritance, 25, 26
intimacy, 34. *see also* same-sex love

Jagendorf, Zvi, 39
James, N.D.G., 9–10
Jhally, Sut, 103
Johnson, Mark, 90, 145n11
Jones, Eldred, 70
Jordan, Constance, 79
Jorden, Edward, 102, 150n7

Kahn, Coppélia, 75
Kerridge, Richard, 149n3
Kiefer, Frederick, 143n17
King, Ynestra, 88
King Lear, 19–32
  analogical thinking in, 25–26
  metaphors of cannibalism in, 24
  misogyny in, 27–28
  nature in, 12–13, 26–27
  weather in, 20–21
  writing of ecophobia in, 5–6
Kleinburg, Seymour, 107
Kline, Stephen, 103
Kolodny, Annette, 153n29
Kristeva, Julia, 83, 140n2

Laird, David, 94
Lakoff, George, 90, 145n11
land, identity and, 23
language, importance of in *The
    Winter's Tale,* 94–95
Laroque, François, 52
Latour, Bruno, 82–83, 123
Lawson, Alan, 83, 103

Leggatt, Alexander, 79
Leiss, William, 103
Les Mots et les choses (Foucault), 92
Levin, Richard, 114, 117
Levy, Neil, 5
Lewin, Jennifer, 114, 121
Lewis, Anthony J., 78
listening/hearing, 46–47
literal/figurative binary, 89–90,
    145n8
Little Ice Age, 20–21, 27, 132n3
Lodge, Thomas, 58
Loomba, Ania, 73
Lowe, Lisa, 35

Macbeth, 13, 102, 118
madness, 99–103, 149n6, 150n10
Maguin, Jean-Marie, 113
Manes, Christopher, 6
manhood, madness and, 103
Manwood, John, 10
mapping of the body, 28, 68–69,
    108
Markley, Robert, 21, 132n3
Marlowe, Christopher, 36
Martyr, Peter (Pietro Martire
    d'Anghiera), 76
Marx, Leo, 49
masculine identity, 21–23, 54, 107,
    138n5
Mazel, David, 50
McLuskie, Kathleen, 7
meat, 3, 54, 138n5, 139nn7–8. *see
    also* vegetarianism
Meeker, Joseph, 113
Mellor, Mary, 88
Merchant, Carolyn, 44, 91–92
Merchant of Venice, The, 105–6,
    152nn20–21
metaphor, 16, 89–92, 145n9,
    145nn11–12, 146n14, 148n26
  animal analogy and, 146n16
  linking women and natural
    world, 93–94
  in *The Winter's Tale,* 92–98
Midland Revolts (1607/8), 33, 37